Computing Supplementum 6

U. Kulisch and H. J. Stetter (eds.)

Scientific Computation with Automatic Result Verification

Springer-Verlag Wien New York

Prof. Dr. Ulrich Kulisch
Institut für Angewandte Mathematik
Universität Karlsruhe
Federal Republic of Germany

Prof. Dr. Hans J. Stetter
Institut für Angewandte
und Numerische Mathematik
Technische Universität Wien
Austria

With 22 Figures

Library of Congress Cataloging in Publication Data.
Scientific computation with automatic result verification / U. Kulisch
  and H. J. Stetter (eds.)
    p. cm. — (Computing. Supplementum ; 6)
Based on papers presented at a conference held Sept. 30–Oct. 2,
1987 in Karlsruhe and co-sponsored by the Institute for Applied
Mathematics of Karlsruhe University and the GAMM Committee on
"Computer Arithmetic and Scientific Computation".
    Includes bibliographies.
    ISBN-13:978-3-211-82063-6 (U.S.)
    1. Numerical calculations-Verification-Congresses.
2. Algorithms-Congresses. I. Kulisch, Ulrich. II. Stetter, Hans
J., 1930– . III. Universität Karlsruhe. Institut für Angewandte
Mathematik. IV. GAMM Committee on "Computer Arithmetic and
Scientific Computation". V. Series: Computing (Springer-Verlag).
Supplementum ; 6.
QA297.S392 1988
519.4—dc19   88-24988
                                                            CIP

ISSN 0344-8029
ISBN-13:978-3-211-82063-6   e-ISBN-13:978-3-7091-6957-5
DOI: 10.1007/978-3-7091-6957-5

Preface

Scientific Computation with Result Verification has been a persevering research topic at the Institute for Applied Mathematics of Karlsruhe University for many years. A good number of meetings have been devoted to this area. The latest of these meetings was held from 30 September to 2 October, 1987, in Karlsruhe; it was co-sponsored by the GAMM Committee on "Computer Arithmetic and Scientific Computation".

This volume combines edited versions of selected papers presented at this conference, including a few which were presented at a similar meeting one year earlier. The selection was made on the basis of relevance to the topic chosen for this volume. All papers are original contributions. In an appendix, we have supplied a short account of the Fortran-SC language which permits the programming of algorithms with result verification in a natural manner.

The editors hope that the publication of this material as a Supplementum of Computing will further stimulate the interest of the scientific community in this important tool for Scientific Computation. In particular, we would like to make application scientists aware of its potential. The papers in the second chapter of this volume should convince them that automatic result verification may help them to design more reliable software for their particular tasks.

We wish to thank all contributors for adapting their manuscripts to the goals of this volume. We are also grateful to the Publisher, Springer-Verlag of Vienna, for an efficient and quick production.

Karlsruhe and Vienna, March 1988   U. Kulisch and H. J. Stetter

Contents

Contents

Appendix

Computing, Suppl. 6, 1–6 (1988)

Automatic Result Verification

U. **Kulisch,** Karlsruhe, and **H.J. Stetter,** Vienna

Abstract — Zusammenfassung

Automatic Result Verification. As an introduction to the following articles, we explain the meaning of automatic result verification as a tool in Scientific Computation; then we shortly sketch its principal methods and put the papers of the volume into a common perspective.

Automatische Ergebnisverifikation. Als Einführung zu den folgenden Artikeln erklären wir zunächst die Bedeutung der automatischen Ergebnisverifikation als einem Werkzeug für das wissenschaftliche Rechnen; dann skizzieren wir kurz die wichtigsten einschlägigen Methoden und setzen die Arbeiten des Bandes in einen übergeordneten Zusammenhang.

1. The Meaning of Result Verification

Scientific Computation is as old as Science and has developed along with it, from Babylonian times until today. A definite milestone along this way was Gauss' computation in 1801 of the orbit of the planetoid Ceres from observations covering only a small portion of the orbit, a computation which permitted the rediscovery of that tiny celestial body after its passage behind the sun. From this event, a direct route leads to the immense computations determining and controlling the satellite orbits which put a man on the moon in 1969.

While an error in Gauss' computations would have remained unnoticed outside a few astronomers' circles, a failure of the moon mission due to erroneous computational control would have been a worldwide spectacle, with human lives at stake. Yet even today, when mathematical modelling and computer simulation are the state of the art throughout all areas of science and, far beyond science, begin to affect our lives, in medical treatment, economical planning, food production etc., the reliability of computed results has remained a tabu: Scientists will argue at length about the validity of their models, yet they will believe the numbers and plots which the computer produces as a consequence of these models.

It is true that everybody admits that digital computers cannot produce accurate results; but it is assumed that — at the validity level of the models — the inaccuracies introduced by the computation do not matter and "they can always be made small enough by using a finer discretization and a higher arithmetic precision".

Examples of computations with no correct digit in the result or of computational artefacts in the outcome of computational simulations have stimulated research on algorithms and arithmetic tools which make the computer display more clearly what it has achieved: Like the old fashioned bookkeeper who would check and recheck his calculations before accepting their results as correct, the computer should verify the accuracy of the piles of numbers and plots which it may discharge so abundantly.

More precisely, "automatic result verification" means that the computer is enabled to produce two values for each result quantity of which it can be rigorously proved that the value of the mathematically defined result quantity lies between them. Naturally, the agreement between these two values defines the accuracy which we can guarantee for our computed results.

Result verification makes it possible to distinguish between effects due to the mathematical model and those due to our computation. Thus it is an indispensible part of safe computer simulation, at least as a tool in the development and testing of reliable production codes. It permits to solve mathematical problems numerically, in a strictly defined sense, rather than to simulate their solution on a digital computer.

When we have stated that result verification generates rigorous inclusions for exact result values, the mathematical problem whose results are enclosed must also be rigorously defined. Each run of a given code with result verification, with given data values, correlates to a specific mathematical problem with specific data: it is this so-called specified problem whose result values are enclosed by the computational results. Only with respect to this specified problem can the assertions about the computation hold.

The introduction of this concept of a specified problem, which may or may not fully coincide with the "given" problem, marks a clear logical borderline between the modelling and computation parts of a simulation. Both parts determine the overall dependability and accuracy; automatic result verification serves to assess the accuracy of the computation reliably, within the computation itself. Thus, automatic result verification reintegrates digital computation into classical mathematics, making it a rigorous tool in the mathematical sense.

2. Methods for Result Verification

Automatic result verification is only possible if a well defined and carefully implemented computer arithmetic is available; see, e.g., [1] or [2]. The IEEE Standard for Binary Floating-Point Arithmetic ([3]) provides a sound basis but has to be supplemented by further arithmetic tools through hardware or software. Program packages like ACRITH ([4]) or ARITHMOS ([5]) have been developed over the last years which provide the necessary arithmetic tools (interval arithmetic, accurate scalar product, accurate standard functions) in the form of subroutines; but this makes the progamming of verification algorithms tedious and intransparent.

The necessary arithmetic tools can be used much easier if they have been integrated into a high-level programming language. Such languages like PASCAL-SC ([6]) and Fortran-SC ([7]) have also been developed over the last years; they have been implemented on various processors from microsystems to mainframes. The availability of these arithmetic and programming tools is required and assumed in all contributions of this volume.

Automatic result verification rests on a number of mathematical and algorithmic techniques which are combined in various ways to achieve the desired rigorous inclusion of the results of the respective specified problem. Naturally, this inclusion must be realistically tight at the same time to provide a result verification which is of practical use. We will shortly sketch the principal techniques in this area:

(i)   A priori error estimation and interval evaluation: Here, the art of the mathematician consists in representing the result by expressions which permit a stable interval evaluation. In this volume, the algorithms for the evaluation of standard functions with dynamic accuracy are examples for the technique.

(ii) Convergent iterative interval methods: A number of classical iteration methods can be put into a form in which they retain their convergence when they are applied to intervals. This permits the iterative reduction of the width of inclusion intervals. Various types of interval defect correction methods (cf. e.g. [8]), like interval Newton methods and their derivatives, are examples of this approach. In this volume, the processes in G. Mayer's paper represent this technique most clearly.

(iii) Fixed point criteria: Special versions of various fixed-point theorems (cf. e.g. [9]) provide the tool for the rigorous verification of a result inclusion, independently of how the inclusion has been obtained. Naturally, the specified problem must be expressible as a fixed-point problem in a suitable way; but this is nearly always possible.

(iv) Staggered corrections: In many situations, it is necessary to generate intermediate results or defects to a dynamically higher accuracy. It is convenient to represent such quantities as sums of a value and (possibly several) corrections each of which is in normal floating-point format (cf. e.g. [10]).

Most of the contributions of this volume use various combinations of several or all of these techniques in order to achieve result verification. Many further clever algorithmic devices are necessary to make the result verification automatic and as failproof as possible.

It should be emphasized that — in a correct routine or package with automatic result verification — a failure can never consist in an incorrect result in the sense that the computed inclusion would not contain the exact result of the mathematical problem (the specified problem). A failure would rather consist in the program's inability to complete its job successfully or in the generation of an inclusion which is unrealistically wide.

3. An Overview of This Volume

The first group of papers is concerned with result verification for standard mathematical problems:

Two papers describe methods for two-point boundary value problems. J. Schröder's algorithm not only generates upper and lower bounds for the solution but also proves the existence of a solution of the analytic problem by a fixed-point approach. J. Weissinger uses a different technique: An interval polynomial is computed which includes the solution of the boundary value problem.

The next two papers consider problems with interval data which may, e.g., represent the uncertainties in the data: Ch. Jansson describes a method which generates upper and lower bounds for the solution set of a linear programming problem with interval data; an automatic sensitivity analysis is also delivered. G. Mayer considers linear systems of equations with interval coefficients; his iterative methods based on matrix splittings converge to an inclusion set of the set of solutions.

Eigenvalue problems are the topic of the next three contributions: G. Alefeld considers systems of quadratic equations, with the generalized eigenvalue problem as a special case. H. Behnke's method, based on Temple quotients and their generalizations, can even handle multiple or clustered eigenvalues. M. Ohsmann describes the verified inclusion of eigenvalues of certain difference equations; his idea may be generalized to the computation of eigenvalues of certain second order differential equations.

The second chapter of this volume contains four contributions which describe the application of result verification techniques in the technical sciences. They treat problems where attempts with traditional algorithmic and arithmetic means have failed to produce reliable answers for quite a while:

A. Ams and W. Klein have developed a software product which computes inclusions of the critical frequences for the bending of rotors. At high rotation speeds, traditional methods used to fail because of an uncontrollable deterioration of the accuracy; here, the accuracy has been improved and verified at the same time.

The paper by D. Cordes is concerned with vibrations of gear drives (eigenvalues of differential equations with periodic coefficient functions). His numerical testing method guarantees the stability of all solutions in the linear case or the boundedness of all perturbed solutions in the nonlinear case.

The problem of finding periodic solutions of differential equations can be transformed into a boundary value problem; hence, automatic result verification may serve as a tool for proving the existence of periodic solutions by computational means. This is the basic idea of the paper by E. Adams, A. Holzmüller, and D. Straub which considers the Oregonator, a well-known model in chemical kinetics which contains ill-conditioned nonlinear stiff differential equations.

A large number of new algorithms for geometric problems have lately become available; however, severe numerical difficulties may occur during their execution. Th. Ottmann, G. Thiemt, and Ch. Ullrich discuss these problems and show, in a model example, how to overcome them by an accurate evaluation of arithmetic expressions with result verification.

The volume ends with five contributions which deal more directly with the tools required in scientific computation with automatic result verification:

A fundamental problem is the reliable evaluation of arithmetic expressions; routines for the evaluation of polynomials and simple expressions with verified last-bit accuracy have been available for some time (see [4] and [5]). R. Lohner now generalizes the polynomial routines to polynomials in several variables while H.C. Fischer, G. Schumacher, and R. Haggenmüller treat expressions in several variables, generalize the techniques to complex and vector-matrix expressions and consider intervals as arguments.

Standard functions (like exp, log, sin, etc.) present a further problem. So far they have been available with last-bit accuracy in packages like ACRITH [4]. If standard functions occur as constituents in other problems in an ill-conditioned setting, it may be necessary to evaluate them more accurately. This requires a set of formulas which permit an evaluation of standard functions (for real and complex point and interval arguments) with a dynamic accuracy. The requested accuracy as well as the base of the floating point system are parameters in the formulas derived by K. Braune and W. Krämer which are based on Taylor series expansions.

The paper by H.J. Stetter which concludes this volume returns to our initial question for the rigorous definition of the mathematical problem whose results are included by the verification process: If the underlying task contains functions as data, the inclusion algorithm must be presented with sufficient information about these functions; otherwise, it cannot decide which is the precise mathematical problem it is supposed to deal with. It turns out that it is necessary to restrict potential data functions to arithmetic expressions and to introduce these into the algorithm as "arithmetic expression strings". Therefore the processing of such data types should be supported by languages for scientific computation.

References

[1] U. Kulisch: Grundlagen des Numerischen Rechnens. Bibliographisches Institut (Reihe Informatik, Nr. *19*), Mannheim/Wien/Zürich 1976.
[2] U. Kulisch, W.L. Miranker: Computer Arithmetic in Theory and Practice. Academic Press, New York, 1981.
U. Kulisch, W.L. Miranker: The Arithmetic of the Digital Computer: A New Approach. SIAM Review *28*, March 1986, 1–40.
[3] IEEE Standard for Binary Floating-Point Arithmetic, ANSI Standard 754–1985.
[4] IBM High-Accuracy Arithmetic Subroutine Library (ACRITH). General Information Manual, GC 33-6163-02, 3rd Edition, April 1986.
IBM High-Accuracy Arithmetic Subroutine Library (ACRITH). Program Description and User's Guide, SC 33-6164-02, 3rd Edition, April 1986.
IBM High-Accuracy Arithmetic Subroutine Library (ACRITH). Reference Summary, GX 33-9009-02, 3rd Edition, April 1986.

[5] ARITHMOS Benutzerhandbuch. SIEMENS AG, Bestellnummer U 2900-J-Z 87-1, Sept. 1986.
BS 2000 ARITHMOS, Kurzbeschreibung, Tabellenheft, SIEMENS AG, Sept. 1986.

[6] U. Kulisch (ed.): PASCAL-SC: A PASCAL Extension for Scientific Computation; information manual and floppy disks; version IBM PC/AT; operating system DOS. B.G. Teubner Verlag (Wiley-Teubner series in computer science), Stuttgart, 1987.
U. Kulisch (ed.): PASCAL-SC: A PASCAL Extension for Scientific Computation; information manual and floppy disks; version ATARI ST. B.G. Teubner Verlag, Stuttgart, 1987.

[7] J.H. Bleher, U. Kulisch, M. Metzger, S.M. Rump, Ch. Ullrich, W. Walter: FORTRAN-SC: A Study of a FORTRAN Extension for Engineering/Scientific Computation with Access to ACRITH. Computing 39, 1987, 93 – 110.

[8] K. Böhmer, P. Hemker, H.J. Stetter: The Defect Correction Approach, in: Defect Correction Methods; Theory and Application, Computing Supplementum, Springer-Verlag, Wien, 1984, 1 – 32.

[9] E. Kaucher, W.L. Miranker: Self-Validating Numerics for Function Space Problems. Academic Press, 1984.

[10] H.J. Stetter: Sequential Defect Correction in High-Accuracy Floating-Point Algorithms, in: Numerical Analysis (Proceedings, Dundee 1983), Lecture Notes in Math., vol. 1066 (1984), 186 – 202.
W. Auzinger, H.J. Stetter: Accurate Arithmetic Results for Decimal Data on Non-Decimal Computers, Computing 35, 1985, 141 – 151.

Prof. Dr. U. Kulisch
Institut für Angewandte Mathematik
Universität Karlsruhe
Englerstrasse 2
D-7500 Karlsruhe
Federal Republic of Germany

Prof. Dr. H.J. Stetter
Institut für Angewandte
und Numerische Mathematik
Technische Universität Wien
Wiedner Hauptstrasse 8–10
A-1040 Wien
Austria

I. Numerical Methods with Result Verification

Computing, Suppl. 6, 9–22 (1988)

Computing
© by Springer-Verlag 1988

A Method for Producing Verified Results for Two-Point Boundary Value Problems

Johann Schröder, Köln

Summary — Zusammenfassung

A Method for Producing Verified Results for Two-Point Boundary Value Problems. This paper presents algorithms for solving second order two-point boundary value problems which yield existence proofs and two-sided bounds (EB-algorithms). The algorithms are composed of two numerical procedures as sub-algorithms, an approximation procedure A for calculating approximate solutions of certain boundary value problems, and an estimation procedure E for calculating bounds of certain continuous functions. Conditions (on the given boundary value problems) are formulated such that the EB-algorithms can be carried out, if these conditions are satisfied and procedures A and E of sufficient accuracy are available.

Ein Verfahren zur verifizierten Lösung von Zweipunkt-Randwertaufgaben. Für Zweipunkt-Randwertaufgaben zweiter Ordnung werden Algorithmen für Existenzbeweis und Einschließung beschrieben (EB-Algorithmen). Ist ein solcher Algorithmus für ein gegebenes Problem durchführbar, so beweist dies die Existenz einer Lösung, welche zwischen den berechneten Schranken liegt. Die EB-Algorithmen setzen sich zusammen aus zwei numerischen Verfahren: einem Verfahren A zur näherungsweisen Lösung von Randwertaufgaben und einem Verfahren E zur Berechnung von Schranken für gewisse stetige Funktionen. Es werden Bedingungen angegeben, unter denen die EB-Algorithmen durchführbar sind, falls Verfahren A und E von hinreichender Genauigkeit zur Verfügung stehen.

1. Introduction

We are interested in boundary value problems with a differential equation

$$-U'' + F[U] = 0 \quad \text{on} \quad [0, 1] \quad , \quad F[U](x) = F(x, U(x), U'(x)) \quad , \qquad (1.1)$$

and linear or nonlinear boundary conditions. Our aim is the construction of numerical algorithms for proving the *existence* of a solution and calculating *bounds* for this solution (briefly: EB-algorithms).

In the present paper we will explain our approach by considering the case of a scalar differential equation with linear boundary conditions. (See, however, the remarks in section 3.3.) Here, we will apply, as theoretical tools, the usual theory of two-sided bounds, and certain generalizations of this theory needed for treating sufficiently general classes of boundary value problems.

The EB-algorithms described here are composed of two procedures as subalgorithms: an approximation procedure A for calculating approximate solutions of certain boundary value problems, and an estimation procedure E for calculating bounds of certain continuous functions (occurring in the EB-algorithm). In procedure **E** rounding errors must be taken into account. It seems hardly possible to construct an EB-algorithm without using such procedures **A** and **E**, or equivalent or stronger tools.

Since the rounding errors in procedure **A** need not be estimated, a series of different "classical" numerical methods for solving boundary value problems could be used, so that one can exploit all the experience gained by applying these methods. For constructing a procedure **E** additional theoretical tools and some type of interval arithmetic are needed. In our program we use the subroutine package ACRITH [5], which is based on the work of Kulisch and others [10].

For our general EB-algorithm we want to have two statements, called (P) and (T), which are independent of the chosen procedures **A** and **E**. Statement (P), to be used in the practical application, has the following form:

(P) *If the* EB-*algorithm can be carried out, then the given boundary value problem has a solution* U^* *which lies within the bounds calculated with the algorithm.*

The theoretical statement (T) shall tell us, under which conditions the EB-algorithm can be carried out. For the case considered here, our final goal is the construction of an EB-algorithm such that a statement of the following form is true.

(T) *Suppose that F is sufficiently smooth, that the given boundary value problem has a solution* U^*, *and that the linearization of the boundary value problem at* U^* *is nonsingular, i.e., described by an invertible linear operator (called* $M'(U^*)$). *Then the* EB-*algorithm for calculating* U^* *can be carried out, if procedures* **A** *and* **E** *of sufficient accuracy are available.*

For explaining our approach we will first discuss in some detail a simple case, where the nonlinearity F does not depend on U', and the usual theory of two-sided bounds with smooth bound functions is applied (section 2). Various generalizations are considered in section 3. Section 4 provides examples.

Notation, and general assumptions:

All inequalities for functions on $[0, 1]$ which will occur are understood to hold pointwise on $[0, 1]$, i.e., $u \leq v \Leftrightarrow u(x) \leq v(x)$ for all $x \in [0, 1]$.

We assume throughout this paper that $F(x, y, p)$ is twice continuously differentiable for $0 \leq x \leq 1$, $y \in \mathbb{R}$, $p \in \mathbb{R}$. (This condition could be relaxed.) The given linear boundary conditions are supposed to have the form

$$B_0[U] = r_0 \quad , \quad B_1[U] = r_1$$

with

$$B_0[U] = -\alpha_0 U'(0) + \gamma_0 U(0) \quad , \quad B_1[U] = \alpha_1 U'(1) + \gamma_1 U(1) \quad ,$$

and constants $\alpha_0, \alpha_1, \gamma_0, \gamma_1, r_0, r_1$ such that

$$\alpha_0 \geq 0 \quad , \quad \gamma_0 = 1 \quad \text{if} \quad \alpha_0 = 0 \quad , \quad \alpha_1 \geq 0 \quad , \quad \gamma_1 = 1 \quad \text{if} \quad \alpha_1 = 0 \quad .$$

Let R denote the space of all $u \in C_2[0, 1]$ such that $B_0[u] = B_1[u] = 0$.

For $v \in C_2[0,1]$ we define an operator $M'(v)$ on R by

$$M'(v)u = -u'' + b[v]u' + c[v]u$$

with continuous functions $b[v]$, $c[v]$ given by

$$b[v](x) = \frac{\partial F}{\partial p}(x, v(x), v'(x)) \quad , \quad c[v](x) = \frac{\partial F}{\partial y}(x, v(x), v'(x)) \quad .$$

This operator is called the *linearization at v of the given differential operator M* ($MU = -U'' + F[U]$). If v satisfies the given boundary conditions, one has for $u \in R$

$$M(v+u) - Mv = M'(v)u + f[u] \quad \text{on} \quad [0,1]$$

with $f[u](x) = f(x, u(x), u'(x))$, and

$$f(x, y, p) = O(y^2 + p^2) \quad \text{for} \quad y, p \to 0 \quad ,$$

uniformly in x. This function f depends also on v. This dependence will at some places be indicated by writing $f(x, y, p; v)$.

An operator $M'(v)$ as above is called *inverse-positive on R*, if $(u \in R, M'(v)u \geq 0)$ $\Rightarrow u \geq 0$.

On the Literature:

The basic inclusion and existence theory used here was provided in [17], for vector-valued problems. The detailed development of a numerical algorithm which provides a rigorous existence proof, as described in this survey, was joint work with M. Göhlen, and M. Plum. In particular, Göhlen and Plum contributed to the work by investigating the treatment of rounding errors, and applying ACRITH. M. Göhlen wrote the programs. A more detailed description of the numerical procedures with a series of modifications will be given in [4]. Results on inverse-positive linear differential operators, and on the theory of two-sided bounds which are applied in this paper can be found in [16].

Other approaches for obtaining existence statements for second order boundary value problems including rounding error estimates are due to Kedem [7], Kaucher and Miranker [6], Lohner and Adams [12], and Lohner [11]. Kedem [7] applies the Kantorovic theorem on Newton procedures in Banach spaces. Kaucher and Miranker [6] use an interval-type fixed-point theorem, assuming that the corresponding operator has a certain contraction property; the second order problems treated in [6] are transformed into a fixed-point equation by introducing the new variable $z = u''$. The papers [12], [11] essentially treat initial value problems; the results can also be used for boundary value problems by applying interval type shooting methods. For further related work see [1], [13], [14].

2. An EB-Algorithm for a Special Case

a) *The given boundary value problem.* We consider a boundary value problem for $U \in C_2[0,1]$ of the form

$$-U'' + F[U] = 0 \quad \text{on} \quad [0,1] \quad , \quad B_0[U] = r_0 \quad , \quad B_1[U] = r_1 \tag{2.1}$$

with $F[U](x) = F(x, U(x))$.

b) *The basic existence and inclusion statement.* Suppose that $\omega \in C_2[0,1]$ is an approximate solution of the given boundary value problem which satisfies the given boundary conditions, and denote its *defect* by $d = d[\omega] = \omega'' - F[\omega]$. Then one obtains for $u = U - \omega$ an equivalent boundary value problem of the form

$$u \in R \quad , \quad L[u] + f[u] = d \quad \text{on} \quad [0,1] \tag{2.2}$$

with $L[u] = -u'' + cu$, $c = c[\omega]$, $f[u](x) = f(x, u(x))$, and $f(x,y) = f(x,y;\omega)$ as in section 1.

Applying the theory of two-sided bounds to this problem, we obtain the following statement.

If functions $\varphi, \psi \in R$ exist such that

$$\varphi \leq \psi \quad , \quad L[\varphi] + f[\varphi] \leq d \leq L[\psi] + f[\psi] \quad , \tag{2.3}$$

then the given boundary value problem has a solution U^* such that $\varphi \leq U^* - \omega \leq \psi$.

c) *Construction of a bound function ψ.* We describe a method for constructing a *bound function* $\psi \in R$ such that φ, ψ with $\varphi = -\psi$ satisfy (2.3). We are mainly interested in "small" bound functions ψ, which will, in general, only exist if the defect d is small. For sufficiently small ψ the nonlinear terms $f[-\psi]$ and $f[\psi]$ will have little influence, so that we may use rough bounds of these terms.

Conditions (2.3) are satisfied for $\psi \in R$ and $\varphi = -\psi$, if $\psi \geq 0$, and

$$|\psi| \leq v \quad , \quad L[\psi] \geq \delta + k \tag{2.4}$$

with constants v, δ, k such that

$$|d| \leq \delta \quad , \tag{2.5}$$

$$|f(x,y)| \leq k \quad \text{for} \quad 0 \leq x \leq 1 \quad , \quad |y| \leq v \quad . \tag{2.6}$$

The inequalities in (2.4), (2.6) are coupled. We solve them in two steps. First a function $\psi_0 \in R$ is calculated such that $L[\psi_0] \approx 1$ on $[0,1]$. The sign $\approx$ indicates that ψ_0 is an approximate solution of the corresponding equation (with $\approx$ replaced by $=$). Then we find a constant $\varepsilon > 0$ such that $\psi = (1 + \varepsilon)\delta\psi_0$ satisfies (2.4), (2.6).

The constant k in (2.6) may be obtained in the following way. One determines a non-decreasing function $G : [0,\infty) \times [0,\infty) \to [0,\infty)$ such that for $v \in C_2[0,1]$:

$$|f(x,y;v)| \leq G(|y|,\tau) \quad \text{for} \quad 0 \leq x \leq 1 \ , \ -\infty < y < \infty \ , \ |v| \leq \tau \ ,$$
$$G(s,\tau) = O(s^2) \quad \text{for} \quad s \to 0 \ , \tag{2.7}$$

and defines $k = G(v, \tau_0)$ with $|\omega| \leq \tau_0$. Such a *majorizing function G* will be used in the following algorithm.

Example: $F(x,y) = r(x) - 25y + y^3$, $f(x,y;v) = y^2(3v(x) + y)$, $G(s,\tau) = s^2(3\tau + s)$.

d) *The algorithm.* Using the results and methods just explained one obtains the following EB-algorithm and statement (P). Here, $\delta, \varepsilon_0, v_0, \delta_0$ denote constants ≥ 0 to be determined, and d, L, and f are defined as above with ω as calculated in step (1). G is a majorizing function as in (2.7). The sign $\approx$ indicates "approximate equality" on $[0, 1]$. $\varepsilon > 0$ denotes some given constant, say $\varepsilon = 0.01$. (See, however, the modifications below.) w is a function defined by

$$w(x) = (x + (1 - x)sgn\alpha_0) \cdot (1 - x + xsgn\alpha_1) \quad .$$

For $u \in R$, $w^{-1}u$ can be extended to a continuous function on $[0, 1]$.

EB-algorithm:

(1) *Solve:* $\omega \in C_2[0, 1]$, $-\omega'' + F[\omega] \approx 0$, $B_0[\omega] = r_0$, $B_1[\omega] = r_1$.

(2) *Solve:* $\psi_0 \in R$, $L[\psi_0] \approx 1$.

(3) *Estimate:* $\psi_0 \geq \varepsilon_0 w$.

(4) *Verify:* $\varepsilon_0 \geq 0$.

(5) *Estimate:* $|\omega| \leq \tau_0$, $|d| \leq \delta$, $\psi_0 \leq v_0$, $|L[\psi_0] - 1| \leq \delta_0$.

(6) *Verify:* $\varepsilon\delta \geq (1 + \varepsilon)\delta\delta_0 + G((1 + \varepsilon)\delta v_0, \tau_0)$.   (2.8)

(P) *If the* EB-*algorithm can be carried out, then the given problem has a solution* U^* *such that* $|U^* - \omega| \leq \psi := (1 + \varepsilon)\delta\psi_0$.

The inequality in step (4) is sufficient for $\psi \geq 0$, the inequality in step (6) sufficient for the differential inequality in (2.4).

In steps (1) and (2) an approximation procedure **A** for solving boundary value problems is needed, in steps (3) and (5) an estimation procedure **E** for obtaining the bounds occurring. In calculating these bounds (and in step (6)) rounding errors must be taken into account. The rounding errors in procedure **A** need not be estimated.

e) *Modifications.* α) For computational reasons one may change the order of the calculations in the EB-algorithm, depending on the chosen procedures **A** and **E**. Also one may use ω such that ω'' is only piecewise continuous.

β) In step (6) one may use a given $\varepsilon > 0$. Another possibility is to consider (2.8) as an inequality for $\eta = \varepsilon\delta$,

$$\eta \geq g(\eta) := (\delta + \eta)\delta_0 + G((\delta + \eta)v_0, \tau_0) \quad ,$$

and solve this inequality by appropriate means. For instance, step (6) may be replaced by

(6'a) *Calculate:* $\eta_0 = g(0)$, $\eta = 2\eta_0$.

(6'b) *Verify:* $\eta \geq g(\eta)$.

In this case, $\psi = (\delta + \eta)\psi_0$.

γ) One need not use a majorizing function G. Step (6) may be replaced by:

(6"a) *Estimate:* $|f(x,y)| \leq k$ for $0 \leq x \leq 1$, $|y| \leq (1+\varepsilon)\delta v_0$.

(6"b) *Verify:* $\varepsilon\delta \geq (1+\varepsilon)\delta\delta_0 + k$.

In this case, the constant τ_0 need not be calculated. The modifications described in β and γ can be combined.

f) *Procedure* **A** *and* **E** *for polynomial differential equations.* Suppose that $F(x,y)$ is a polynomial in x and y. Then, if ω and ψ_0 are polynomials (or piecewise polynomial functions) all functions to be estimated in steps (3) and (5) are polynomials (or piecewise polynomial functions). For this case we developed an estimation procedure **E**, based on the following result of Ehlich and Zeller [2]:

If P_N is a polynomial of degree $\leq N$, then for $|x| \leq 1$

$$|P_N(x)| \leq C(N,m) \max\{P_N(t_j) : j = 1, 2, \dots, m\} \tag{2.9}$$

with $C(N,m) = (\cos\frac{\pi N}{2m})^{-1}$ *(for instance,* $C(N,2N) = \sqrt{2}$ *), and* $t_j = \cos\frac{2j-1}{2m}\pi$ *(the zeros of the* m*-th Chebyshev polynomial).*

As procedure **A** we used a collocation method with polynomials and Chebyshev collocation points (zeros or extreme points of Chebychev polynomials transformed to $[0,1]$), or piecewise polynomial functions and corresponding collocation points. For nonlinear problems the collocation method was combined with a Newton iterative procedure. An appropriate method for calculating a start function for the iteration belongs also to procedure **A**.

We cannot discuss here all details of these procedures. We will, however, very briefly describe one difficulty which we encountered. Suppose, for simplicity, that $B_0[u] = u(0)$, $B_1[u] = u(1)$, and that polynomials are used in the collocation method. Then ω can be written in the form

$$\omega(x) = r_0(1-x) + r_1 x + \sum_{i=2}^{N} \alpha_i \varphi_i(x) \tag{2.10}$$

with a polynomial φ_i of degree i such that $\varphi_i(0) = \varphi_i(1) = 0$.

A reasonable choice is $\varphi_i(x) = \int_0^x L_{i-1}(2t-1)dt$ with L_i denoting the i-th Legendre polynomial. The values $\varphi_i(x_j)$, $\varphi_i'(x_j)$, $\varphi_i''(x_j)$ at the collocation points x_j, which are needed in procedure **A**, can be calculated by applying simple recursion formulas with respect to i. In procedure **E** one would like to use an interval arithmetical version of these formulas, or some other interval arithmetical algorithm related to these formulas. All such algorithms which we tested, however, proved to be numerically unstable. It may be possible to combine the interval arithmetical recursion formulas with some sophisticated method of defect correction (based on an idea developed in [15]). This remains to be investigated.

Instead, we use now the functions $\varphi_i(x) = x(1-x)T_{i-2}(2x-1)$ with T_i denoting the i-th Chebyshev polynomial. The values to be calculated then are linear combinations of the terms $\cos\frac{i\pi}{N}$, and $\sin\frac{i\pi}{N}$, which can be computed with interval-arithmetical algorithms for the sin- and cos-function (available, e.g., in ACRITH).

g) *Other procedures* **A** *and* **E**. In constructing a procedure **A** one could use a series of different methods for calculating approximate solutions of boundary value problems. Procedure **A** must provide functions defined on the entire interval $[0, 1]$. Nevertheless, one could also make use of difference methods, or other methods which yield only values in finitely many points t_i (e.g., equidistant t_i). One possibility would be to interpolate these values. We intend, however, to investigate the following procedure.

Given approximate values at points t_i, one applies a Newton-collocation method with piecewise polynomial functions. In the first step one uses the points t_i as collocation points, and obtains an approximate solution defined on $[0, 1]$. In order to improve this approximation one then continues the Newton-collocation method with collocation points as described under f). This procedure starts with approximate values (in the points t_i), which can be computed with any method suitable for the particular boundary value problem.

Constructing a procedure **E** means solving the basic problem of calculating bounds for (certain classes of) continuous functions. For polynomials (or piece-wise polynomial functions) formula (2.9) is a suitable tool. U. Gärtel [3] provided corresponding tools for more general functions. She considered functions g such that $g(x)$ can be obtained by finitely many applications of "elementary" functions (such as $+, -, \cdot, :, \sin, \cos, \log, \cdots$), and provided an algorithm for calculating $m \in \mathbb{N}$, and a constant $K(m)$ such that for $|x| \leq 1$ and a given $r > 0$:

$$|g(x)| \leq K(m) \max\{g(x_j) : j = 1, 2, \cdots, m\} + (K(m) + 1)r \quad .$$

h) *On the theory of the* EB-*algorithm*. Suppose that the given boundary value problem has a solution U^*. One would like to know, under which conditions the EB-algorithm for calculating U^* can be carried out, i.e., under which conditions one can *prove* the existence of U^* by applying the EB-algorithm. (Of course, these conditions cannot, and need not be verified before applying the EB-algorithm. For the practical application one uses statement (P).)

The accuracy of an approximation of a solution will be measured using the norm $\| u \|_2 = \|u\|_\infty + \|u'\|_\infty + \|u''\|_\infty$ with $\| \; \|_\infty$ denoting the maximum norm. Thus, the defect of an approximation ω of U^* becomes arbitrarily small, if ω is a sufficiently good (sufficiently accurate) approximation of U^*, i.e., if $\| U^* - \omega \|_2 \leq \varepsilon_1$ with a sufficiently small $\varepsilon_1 > 0$.

For the above EB-algorithm the following statement can be proved.

(T') *Suppose that the given boundary value problem has a solution* U^*, *and that* $M'(U^*)$ *is inverse-positive on R. Then the* EB-*algorithm can be carried out, if procedures* **A** *and* **E** *of sufficient accuracy are available.*

Let us briefly indicate, in which way this statement can be proved. If $M'(U^*)$ is inverse-positive on R, then there is a $z \geq 0$ in R such that $M'(U^*)z \equiv 1$. This function z then is *strictly positive*, i.e., $w^{-1}z \geq$ const. > 0. If ω is a sufficiently good approximation of U^*, then $M'(\omega)z \geq$ const. > 0. Therefore, $M'(\omega)$ is also inverse-positive on R, and hence a strictly positive $z_0 \in R$ exists such that $M'(\omega)z_0 \equiv 1$. If, in addition, ψ_0 is a sufficiently good approximation of z_0, then ψ_0 is strictly positive, so that step (4) can be carried out (if the lower bound ε_0 calculated in step (3) is sufficiently good).

Moreover, there are bounds $\tilde{\tau}_0 > 0$ and $\tilde{v}_0 > 0$ such that $|\omega| \leq \tilde{\tau}_0$ and $|\psi_0| \leq \tilde{v}_0$ for all sufficiently good approximations ω and ψ_0. For such functions ω, ψ_0 the inequality

$$\varepsilon \delta \geq (1 + \varepsilon) \delta \delta_0 + G((1 + \varepsilon) \delta 2\tilde{v}_0, 2\tilde{\tau}_0)$$

is sufficient for (2.8) (if the bounds v_0, τ_0 are sufficiently good, so that at least $v_0 \leq 2\tilde{v}_0$, $\tau_0 \leq 2\tilde{\tau}_0$). Because of the second property in (2.7), the above inequality holds (for a fixed $\varepsilon > 0$), if δ and δ_0 are sufficiently small, i.e., if ω and ψ_0 are sufficiently good approximations (and if the bounds δ, δ_0 are sufficiently good).

3. More General Problems and Methods

We want to generalize the EB-algorithm in section 2 in such a way that

a) differential equations of the general form (1.1) can be treated (where F may depend on U');

b) the inverse-positivity requirement in (T') is replaced by a weaker condition;

c) the generalized EB-algorithm is also composed of the procedures **A** and **E**;

d) the relations to be verified in the generalized EB-algorithm have a form as simple as those in the EB-algorithm of section 2.

Throughout this section we consider boundary value problems for $U \in C_2[0, 1]$ of the form

$$-U'' + F[U] = 0 \quad \text{on} \quad [0, 1] , \quad B_0[U] = r_0 , \quad B_1[U] = r_1 ,$$
$$F[U](x) = F(x, U(x), U'(x)) \quad . \tag{3.1}$$

If $\omega \in C_2[0, 1]$ is a given approximate solution which satisfies the given boundary conditions, we obtain for $u = U - \omega$ the equivalent problem

$$u \in R \quad , \quad L[u] + f[u] = d \quad \text{on} \quad [0, 1] \quad , \tag{3.2}$$

with $\quad L[u] = -u'' + bu' + cu$, $b = b[\omega]$, $c = c[\omega]$, $d = d[\omega]$, $f[u](x) = f(x, u(x), u'(x); \omega)$.

In section 3.1 we use smooth bound functions ψ ($\in C_2[0, 1]$). In sections 3.2 and 3.3 bound functions ψ with *breakpoints* ξ are considered, i.e., functions ψ such that $\psi'(\xi + 0) \neq \psi'(\xi - 0)$ at the breakpoints. The corresponding EB-algorithms can be carried out under weaker conditions on the given problem.

3.1 Estimation of the Derivative

In order to obtain an inclusion statement $\varphi \leq u \leq \psi$, one tries to prove also an inclusion $\Phi \leq u' \leq \Psi$ of the derivative. For our purpose, it suffices to consider estimates $|u| \leq \psi$, $|u'| \leq \kappa$ with a constant $\kappa > 0$.

(A) *If $\psi \in C_2[0, 1]$ and $\kappa \in \mathbb{R}$ satisfy $\psi \geq 0$, $\kappa > 0$, $|\psi'| \leq \mu \leq \kappa$, and the inequalities (2.4) with constants v, μ, δ, k such that (2.5) holds and*

$$|f(x, y, p)| \leq k \quad \text{for} \quad 0 \leq x \leq 1 \quad , \quad |y| \leq v \quad , \quad |p| \leq \mu \quad , \tag{3.3}$$

then there is a $u \in R$ such that

$$L[u] + \widehat{f}[u] = d \quad on \quad [0,1] \quad , \quad and \quad |u| \leq \psi \tag{3.4}$$

with $\widehat{f}[u](x) = f(x, u(x), \widehat{u}'(x))$, $\quad \widehat{u}'(x) = \max\{-\kappa, \min\{u'(x), \kappa\}\}$.

(B) *If, in addition,*

$$|b| \leq \beta \quad , \quad |c| \leq \gamma \quad , \quad \psi(0) + \psi(1) \leq \Lambda \quad ,$$
$$|f(x, y, p)| \leq k_0 \quad for \quad 0 \leq x \leq 1 \quad , \quad |y| \leq v \quad , \quad |p| \leq \kappa \quad , \tag{3.5}$$

$$\kappa^2 - 4\beta v \kappa \geq 4v(\gamma v + \delta + k_0) + \Lambda^2 \quad , \tag{3.6}$$

with constants $\beta, \gamma, \Lambda, k_0$, then the relations (3.4) imply $|u'| \leq \kappa$. Consequently, the given boundary value problem has a solution U^* such that $|U^* - \omega| \leq \psi$, $|(U^* - \omega)'| \leq \kappa$.

Statement (A) follows from the theory of two-sided bounds. Let us here only provide a simple proof of statement (B), which applies arguments used in the theory of Nagumo conditions.

If $|u'| \leq \kappa$ does not hold, there is a t with $\kappa < |u'(t)| =: \overline{\kappa}$. Let $u'(t) = \overline{\kappa}$ (the case $u'(t) = -\overline{\kappa}$ can be treated analogously). Then there is an s such that $u'(s) \leq \Lambda$ ($\leq \kappa < \overline{\kappa}$). Let $s < t$ (the case $s > t$ can be treated analogously). Without loss of generality we may also assume that $0 \leq \Lambda < u'(x) < \overline{\kappa}$ for $s < x < t$. Using (3.4) and the inequalities required we derive

$$u''(x) \leq K := \beta\overline{\kappa} + \gamma v + k_0 + \delta \quad for \quad s \leq x \leq t \quad .$$

Multiplying this inequality by $u'(x)$ and integrating over $[s, t]$ we obtain $\frac{1}{2}(\overline{\kappa}^2 - \Lambda^2) \leq K(u(t) - u(s)) \leq 2Kv$. This inequality contradicts (3.6).

For calculating ψ and κ we proceed in a way analogous to that in section 2. We first determine suitable quantities ψ_0 and κ_0, and then use $\psi = (1 + \varepsilon)\delta\psi_0$, $\kappa = (1 + \varepsilon)\delta\kappa_0$. Here κ_0 is chosen in such a way that $\widetilde{\kappa} = \delta\kappa_0$ satisfies $\widetilde{\kappa}^2 - 4\beta v\widetilde{\kappa} = 4\gamma v^2 + \Lambda^2$.

The constants k and k_0 in (3.3), (3.5) can be obtained with the aid of a (non-decreasing) majorizing function $G : [0, \infty)^4 \to [0, \infty)$, which is required to have the following properties:

$$|f(x, y, p; v)| \leq G(|y|, |p|, \tau_0, \tau_1)$$
$$for \quad 0 \leq x \leq 1 \ , \ y \in \mathbb{R} \ , \ p \in \mathbb{R} \ , \ |v| \leq \tau_0 \ , \ |v'| \leq \tau_1 \quad , \tag{3.7}$$
$$G(s, t, \tau_0, \tau_1) = O(s^2 + t^2) \quad for \quad s, t \to 0 \quad .$$

Using the approach described one can construct an EB-algorithm for which statements (P) and (T') hold also. This algorithm can be obtained by simplifying the algorithm in section 3.2 in a way described there.

3.2 Bound Functions with One Breakpoint

The crucial condition in the EB-algorithm above is the requirement $\varepsilon_0 \geq 0$, which yields $\psi_0 \geq 0$, and hence $\psi \geq 0$. In a concrete case it may be impossible to construct a smooth function $\psi_0 \geq 0$ as required in the algorithm. But it is always possible (at least theoretically) to construct an analogous functions ψ_0 with breakpoints. In the next section we will make some comments on the choice of these breakpoints. Here we will first discuss the case where (only) one breakpoint ξ is given, in order to explain the basic ideas.

Let $\xi \in (0, 1)$. Then define R_ξ to be the space of all $u \in C_0[0, 1]$ such that the restrictions of u to the intervals $[0, \xi]$ and $[\xi, 1]$ are twice continuously differentiable, and $B_0[u] = B_1[u] = 0$. For $u \in R_\xi$ the term $L[u](\xi)$ shall be identified with $L[u](\xi - 0)$ (or, alternatively, $L[u](\xi + 0)$).

We will replace the differential equation in (3.2) at the point ξ by some type of "boundary condition" $u(\xi) = \ldots$. Then we essentially have to deal with two boundary value problems, one on the interval $[0, \xi]$, and one on $[\xi, 1]$, and the bound function ψ need not be smooth at ξ. (These two boundary value problems are coupled only in a rather weak way by the condition at ξ.) It turns out that for the theory of the corresponding EB-algorithm not the inverse-positivity of $M'(U^*)$ is important, but the inverse-positivity of the linearizations at U^* which belong to the two boundary value problems mentioned. We will here exploit the fact that the inverse-positivity of a differential operator depends on the length of the given interval (without giving the precise meaning of this statement).

For constructing a "boundary condition" at ξ we calculate a *breakpoint function* $g \in R_\xi$ such that $g(\xi) = 1$, and $L[g] \approx 0$, i.e., $L[g] \approx 0$ on $[0, \xi]$, and $L[g] \approx 0$ on $[\xi, 1]$ for the corresponding restrictions of g.

Given such a function g, we obtain by partial integration for $u \in R$

$$\int_0^1 pgL[u]\, dx = \sigma u(\xi) + \int_0^1 pL[g]u\, dx$$

with $\sigma = g'(\xi - 0) - g'(\xi + 0)$, $p(x) = \exp\{-\int_\xi^x b(t)\, dt\}$. Thus, if $\sigma \neq 0$, and if u is a solution of (3.2), we have

$$u(\xi) = \sigma^{-1} \int_0^1 p\{g(d - f[u]) - L[g]u\}\, dx \quad .$$

Replacing the condition $L[u](\xi) + f[u](\xi) = 0$ in the boundary value problem (3.2) (for $u \in R$) by the above relation for $u(\xi)$, we obtain a *transformed boundary value problem for $u \in R_\xi$*.

We will not discuss here the theory of this transformation (conditions for $\sigma \neq 0$, equivalence statements etc.).

The theory of two-sided bounds for boundary value problems of the usual form can be carried over to more general problems, such as the transformed problem. Since f may depend on u', one has also to use the method of modification applied in section 3.1 (where u' is replaced by $\hat{u}'$). Instead of a differential inequality at ξ, one then obtains an integral inequality, which is satisfied, if (with k_0 in (3.5))

$$\psi(\xi) \geq |\sigma|^{-1} \int_0^1 p\{|g|(|d| + k_0) + |L[g]|\psi\}\, dx \quad . \tag{3.8}$$

Using these theories one obtains the following EB-algorithm (which, of course, can be applied without knowing all these theories). Here, a majorizing function G as in (3.7) is used. The algorithm may be modified in a way analogous to that explained in section 2e.

EB-algorithm:

(1) *Solve:* $\omega \in C_2[0,1]$, $-\omega'' + F[\omega] \approx 0$, $B_0[\omega] = r_0$, $B_1[\omega] = r_1$.

(2) *Solve:* $g \in R_\xi$, $L[g] \approx 0$, $g(\xi) = 1$.

(3) *Verify:* $\sigma := g'(\xi - 0) - g'(\xi + 0) \neq 0$.

(4) *Estimate:* $|g| \leq \rho$, $|L[g]| \leq \Delta$, $|b| \leq \beta$, $|c| \leq \gamma$.

(5) *Solve:* $\psi_0 \in R_\xi$, $L[\psi_0] \approx 1$, $\psi_0(\xi) = q$

with $q = q_\xi = |\sigma|^{-1}\rho \exp[\beta \max(\xi, 1 - \xi)]$.

(6) *Estimate:* $w^{-1}\psi_0 \geq \varepsilon_0$.

(7) *Verify:* $\varepsilon_0 \geq 0$.

(8) *Estimate:* $|\omega| \leq \tau_0$, $|\omega'| \leq \tau_1$, $|d| \leq \delta$, $|\psi_0| \leq v_0$,

$|\psi_0'| \leq \mu_0$, $|L[\psi_0] - 1| \leq \delta_0$.

(9) *Calculate:* $\zeta_0 = \frac{1}{2}(\psi_0(0) + \psi_0(1))$, $\Delta_0 = v_0\rho^{-1}\Delta$,

$\chi_0 = v_0\beta + [\zeta_0^2 + v_0^2(\beta^2 + \gamma) + v_0(1 - \Delta_0)]^{\frac{1}{2}}$,

$\kappa_0 = \max(\mu_0, 2\chi_0)$.

(10) *Verify:* $\varepsilon\delta \geq (1 + \varepsilon)\delta\delta_0 + G((1 + \varepsilon)\delta v_0, (1 + \varepsilon)\delta \mu_0, \tau_0, \tau_1)$,

$\varepsilon\delta \geq (1 + \varepsilon)\delta\Delta_0 + G((1 + \varepsilon)\delta v_0, (1 + \varepsilon)\delta \kappa_0, \tau_0, \tau_1)$.

For this algorithm statement (P) holds also. Moreover, we have

(T") *Suppose that the given problem* (3.1) *has a solution* U^*, *that* $M'(U^*)$ *is invertible on* R, *and that the operator* A *defined as follows is inverse-positive on* R_ξ:

$$Au(x) = M'(U^*)(x) \quad \text{for} \quad x \neq \xi \quad , \quad Au(\xi) = u(\xi) \quad .$$

Then the EB-algorithm can be carried out, if procedures **A** *and* **E** *of sufficient accuracy are available.*

The first inequality in step (10) is sufficient for the differential inequality for ψ. If the second inequality in step (10) holds, then *both* the inequality (3.6), and the *breakpoint condition* (3.8) are satisfied.

The above algorithm can also be applied, when a smooth function ψ shall be calculated (as discussed in section 3.1). In this case, steps (2), (3), (4), and the condition for $\psi_0(\xi)$ in step (5) are omitted. Also, one requires $\psi_0 \in R$, and defines $\Delta_0 = 0$.

Remark: In [8], [9] Küpper used also non-smooth two-sided bounds. Applying the theory of Range-Domain-Implications he proved inclusion theorems for linear scalar boundary value problems, and used these results in treating corresponding nonlinear problems with nonlinearity $F(x, U)$. Küpper already worked with auxiliary functions which correspond to our breakpoint functions.

3.3 On the General EB-Algorithm

When several breakpoints are used, one needs a breakpoint function $g_\xi \in R_\xi$ for each breakpoint ξ. In this case one determines $g = g_\xi$ in such a way that $L[g] \approx 0$ and $g'(\xi - 0) - g'(\xi + 0) \approx 1$. Now the function ψ_0 in step (5) may have a break at each breakpoint ξ, and ψ_0 must satisfy $\psi_0(\xi) = q_\xi$ at each ξ. Furthermore, Δ_0 in step (9) has to be defined somewhat differently. In the corresponding theoretical statement one requires the inverse-positivity of an operator A such that $Au(\xi) = u(\xi)$ for each breakpoint ξ, and $Au(x) = M'(U^*)(x)$, if x is not a breakpoint.

The question arises, in which way one can choose the breakpoints. In many cases the choice is not very difficult (see the examples below). Nevertheless, one would like to have an algorithm for determining the breakpoints (and for deciding whether breakpoints are needed or not). There are several possibilities for constructing such an algorithm. One will exploit one of the many properties of inverse-positive linear differential operators (e.g., oscillation properties). Here we cannot discuss this in detail. In order to show, however, that such an algorithm can be constructed, let us describe one method which can be explained rather easily.

One starts the EB-algorithm with no breakpoint. If the number ε_0 computed is positive, one continues (with no breakpoint). Otherwise, one introduces one breakpoint (say $\xi = \frac{1}{2}$), starts again (using now the algorithm in section 3.2), and checks whether ε_0 is positive. In this way one continues, introducing more and more suitable breakpoints, until the corresponding number ε_0 is positive. For instance, one may use equidistant breakpoints.

Of course, this is only a rough description. One has also to consider the computing accuracy in the various steps. But in this way one can obtain an algorithm such that statement (T) holds, as formulated in the introduction.

Similar algorithms as described here can also be developed for problems with nonlinear boundary conditions, and for functions ω which satisfy these boundary conditions only approximately (see [4]). The nonlinearities in the differential equation, and the boundary conditions may also contain some functional terms, such as integrals.

4. Examples

We consider the problem

$$-U'' - 25U + U^3 = \lambda x(1-x) \quad , \quad U(0) = U(1) = 0 \tag{4.1}$$

with a real λ. This problem has one solution U_λ, if $|\lambda| > \lambda_0 \approx 102.29$, two solutions for $|\lambda| = \lambda_0$, and three solutions for $|\lambda| < \lambda_0$. The values $(\lambda, U_\lambda(\frac{1}{2}))$ for all these solutions constitute an S-shaped curve, which is antisymmetric with respect to the origin. It consists of three branches, I, II, and III.

We applied the EB-algorithm in section 3.2, and its simplification for the case of no breakpoint, thus obtaining bounds for the error $U_\lambda - \omega$, and for its derivative. For the branches I and III no breakpoint was used, for branch II one breakpoint $\xi = \frac{1}{2}$. At the endpoints of branch II (where $|\lambda| = \lambda_0$) the EB-algorithm cannot be applied, since $M'(U^*)$ is not invertible.

Some results are given in the table. They are obtained with approximation and estimation procedures as discussed in section 2f. For branches I and III we used polynomials, for branch II functions which are polynomials on each of the intervals $[0, \frac{1}{2}]$ and $[\frac{1}{2}, 1]$. N denotes the order of the polynomials used in calculating ω , N_0 the order for g and ψ_0 . The table provides rounded values of $\omega(\frac{1}{2})$ and $\omega'(0)$, in order to indicate the size of these terms, and *error exponents* v_E and κ_E. This are natural numbers such that $\psi \leq 10^{-v_E}, \kappa \leq 10^{-\kappa_E}$. For branch II the case $(N, N_0) = (40, 20)$ was not treated.

In [4] further results on this problem will be given, and further examples will be treated, such as the problem

$$-U'' - \lambda U [1 - \tfrac{1}{2}(U')^2 - \tfrac{1}{8}(U')^4] = 0 \quad , \quad U(0) = U'(1) = 0 \quad .$$

This problem has infinitely many branches of solutions, which bifurcate from the trivial solution $U \equiv 0$ at the eigenvalues $\lambda_k = \tfrac{1}{4}(2k-1)^2\pi^2 \quad (k = 1, 2, ...)$ of the linearized problem. We applied the EB-algorithm to the first and second branch, using no breakpoint for the first branch, and one breakpoint $\xi = \frac{2}{3}$ for the second. For the k-th branch one will try to work with $k-1$ breakpoints.

Branch	λ	$\omega(\frac{1}{2})$	$\omega'(0)$	$(N, N_0) = (20, 10)$		$(N, N_0) = (30, 10)$		$(N, N_0) = (40, 20)$	
				v_E	κ_E	v_E	κ_E	v_E	κ_E
	0	-4.4	17	5	4	8	7	12	11
	50	-3.9	15	5	3	10	9	12	11
I	100	-2.9	10	4	2	9	8	12	11
	102	-2.7	9.2			9	8	11	10
	102.2	-2.6	9.0					11	10
	102.28	-2.598	8.8					10	9
	0	0	0						
	50	-0.9	2.7	12	11	12	11		
II	100	-2.3	7.5	9	8	11	11		
	102	-2.5	8.3	9	8	11	10		
	102.2	-2.5	8.5	8	7	11	10		
	102.28	-2.578	8.7	7	6	10	9		
	0	4.4	17	5	4	8	7	12	11
	50	4.7	19	4	3	8	6	12	10
III	100	5.0	21	4	3	7	6	12	11
	1000	7.4	37	3	2	6	4	9	8
	10000	14	99	2	0	4	2	7	5

Table: Bounds for solutions U_λ of example (4.1):

$$|U_\lambda - \omega| \leq \psi \leq 10^{-v_E} \quad , \quad |(U_\lambda - \omega)'| \leq \kappa \leq 10^{-\kappa_E} \quad .$$

References

1. Cruickshank, D.M., Wright, K.: Computable Error Bounds for Polynomial Collocation Methods. SIAM J. Numer. Anal. 134-151 (1978).

2. Ehlich, H., Zeller, K.: Schwankung von Polynomen zwischen Gitterpunkten. Math. Z. 86, 41-44 (1964).

3. Gärtel, U.: Fehlerabschätzungen für vektorwertige Randwertaufgaben zweiter Ordnung, insbesondere für Probleme aus der chemischen Reaktions-Diffusions-Theorie. Dissertation Köln 1987.

4. Göhlen, M., Plum, M., Schröder, J.: A numerical algorithm for existence proofs for two-point boundary value problems. In preparation.

5. IBM-Subroutine Library: High Accuracy Arithmetic, ACRITH.

6. Kaucher, E.W., Miranker, W.L.: Self-Validating Numerics for Function Space Problems. Academic Press, New York 1984.

7. Kedem, G.: A Posteriori Error Bounds for Two-Point Boundary Value Problems. SIAM J. Numer. Anal. 18, 431-448 (1981).

8. Küpper, T.: Einschließungsaussagen für gewöhnliche Differentialoperatoren. Numer. Math. 25, 201-214 (1976).

9. Küpper, T.: Einschließungsaussagen bei Differentialoperatoren zweiter Ordnung durch punktweise Ungleichungen. Numer. Math. 30, 93-101 (1978).

10. Kulisch, U.W., Miranker, W.L.: Computer Arithmetic in Theory and Practice. Academic Press, New York 1981.

11. Lohner, R.: Enclosing the Solution of Ordinary Initial- and Boundary-Value Problems. In: Proc. 11th IMACS World Congress, pp. 99-102, Oslo 1985.

12. Lohner, R., Adams, E.: Einschließen der Lösung gewöhnlicher Anfangs- und Randwertaufgaben. ZAMM 64, T295-T297 (1984).

13. McCarthy, M.A., Tapia, R.A.: Computable a posteriori L_∞-Error Bounds for the Approximate Solution of Two-Point Boundary Value Problems. SIAM J. Numer. Anal. 12, 919-937 (1975).

14. Nickel, K.: The Construction of a priori Bounds for the Solution of a Two-Point Boundary Value Problem with Finite Elements I. Computing 23, 247-265 (1979).

15. Ruttmann, B.: Untersuchungen zur Fehlerabschätzung von polynomialen Näherungslösungen bei Randwertaufgaben gewöhnlicher Differentialgleichungen. Dissertation Universität Köln 1982.

16. Schröder, J.: Operator Inequalities. Academic Press 1980.

17. Schröder, J.: Existence proofs for boundary value problems by numerical algorithms. Report Univ. Cologne 1986.

Dr. J. Schröder
Mathematisches Institut
Universität Köln
D-5000 Köln 41
Federal Republic of Germany

Computing, Suppl. 6, 23–32 (1988)

A Kind of Difference Methods for Enclosing Solutions of Ordinary Linear Boundary Value Problems

Johannes Weissinger, Karlsruhe

Summary — Zusammenfassung

A Kind of Difference Methods for Enclosing Solutions of Ordinary Linear Boundary Value Problems.
In [5] a method is described, how to build algorithms for enclosing the solution $u(x)$ of ordinary linear differential problems by an interval polynomial $Y(x)$. In this paper the method is used for enclosing the values $u(x_i)$ of boundary value problems at equidistant points x_i by intervals Y_i. The Y_i are determined by a system of linear equations which, for simple problems, is tridiagonal and can be solved easily.

Eine Art Differenzenverfahren zum Einschließen der Lösungen gewöhnlicher, linearer Randwertprobleme. In [5] ist eine Methode beschrieben, wie Algorithmen konstruiert werden können, um die Lösung $u(x)$ von gewöhnlichen linearen Differentialgleichungsproblemen durch ein Intervallpolynom $Y(x)$ einzuschließen. In dieser Arbeit wird die Methode verwendet, um die Werte $u(x_i)$ bei Randwertproblemen in äquidistanten Punkten x_i durch Intervalle Y_i einzuschließen. Die Y_i werden durch ein lineares Gleichungssystem bestimmt, das für einfache Probleme tridiagonal ist und leicht gelöst werden kann.

1. Introduction

In [5] a method was described how to construct algorithms for enclosing the solution of initial and boundary value problems of ordinary linear differential equations. If one assumes that the coefficients of the problem are polynomials and if one puts a polynomial "Ansatz" of degree n into the differential and conditional equations, then the number of equations, which result from a comparison of coefficients, is greater than the number $n+1$ of unknown coefficients in the Ansatz. By enclosing some of the higher powers by interval-polynomials of smaller degree the total degree can be reduced such that the comparison of coefficients yields as many equations as unknowns. But these equations are now interval equations and the solution is an interval vector $(Y_0, Y_1, \ldots, Y_n)^T$. Then under certain conditions, the interval polynomial

$$Y(x) := \sum_{i=0}^{n} Y_i x^i$$

includes the solution $u(x)$ of the differential problem: $u(x) \in Y(x)$.

If some functions of the problem are not polynomials, they have to be enclosed by polynomials. Then the procedure is the same as above. Obviously, in order to get a sharp enclosure of the solutions $u(x)$, these functions must be enclosed sharply and for this the problem functions must be sufficiently smooth.

And even if all functions are smooth, there may arise difficulties with the evaluation of $Y(x)$ for large values of x, if the "approximation degree" n is large, i.e. if the interval of the problem is large.

All these difficulties disappear, if we can split the whole interval into small ones and approximate resp. enclose separately on each of these small intervals. For initial value problems this principle was discussed already in [5]. In this paper we apply it to boundary value problems.

For a better understanding of the arguments presented in the next sections the following remarks may be useful.

1. For the theory in [5] to be valid two assumptions have been made there. The first one is that the boundary value problem satisfies a maximum principle or has the property that certain inequalities in the problem imply inequalities for the solution. The second one claims that at most two components of the solution of a certain system of linear interval equations contain zero as inner point. These assumptions always should be valid also in the sequel.

2. The enclosure method described in [5] which is also the fundament of the method to be developped in the sequel is based on a lemma for systems of linear interval equations. Usually (cf. [1], p. 179) and also in this lemma such a system is only a short notation for a set of point systems and one is interested in the set S of all point vector solutions of these systems. An interval vector V is a solution of the interval system if every point vector of S is contained in V. In order to determine an interval solution any change in the equations that does not diminish the set S of point solutions is allowed. Especially, every operation that is used in solving point systems like adding multiples of equations to other equations is allowed though it may be against the rules of interval arithmetic.

2. Enclosure in the Midpoint of the Basic Interval

By the methods described in [5] or elsewhere (e.g. [2]) we can obtain an inclusion

$$u(x) \in Y(x) := \sum_{i=0}^{n} Y_i x^i, \quad x \in X \tag{2.1}$$

of the solution $u(x)$ by a polynomial $Y(x)$ with interval coefficients for all x of the basic interval X.

Assuming $0 \in X$ we especially have

$$u(0) \in Y_0. \tag{2.2}$$

Enclosing $u(x)$ for $x=0$ only yields two advantages:

1.) It is not necessary to compute all coefficients $Y_i (i=0, 1, \ldots, n)$ from the system of linear equations. Sometimes, e.g. in the model problem (2.3), Y_0 can be determined explicitly.

2.) The evaluation of the polynomial in (2.1), which may be very ill conditioned, is not necessary.

First, we assume that $X=[-1, 1]$ and $x=0$ is the midpoint of X. By a linear transformation, which does not change the problem essentially, this can always be obtained.

For describing the procedure we use the following model problem

$$Lu(x):=u''(x)-au(x) \overset{!}{=} g(x), \quad x \in X:=[-1, 1] \tag{2.3a}$$

$$u(-1) \overset{!}{=} \alpha, \quad u(1) \overset{!}{=} \beta \tag{2.3b}$$

where a, α, β are given real constants. The given function $g(x)$ shall be enclosed by an interval polynomial $G(x)$:

$$g(x) \in G(x):= \sum_{i=0}^{n-2} G_i x^i, \quad x \in X. \tag{2.4}$$

For reasons of simplicity the integer n shall be even and ≥ 4.

For the Ansatz polynomial

$$y(x):= \sum_{i=0}^{n} y_i x^i, \quad y_i \in \mathbb{R} \tag{2.5}$$

we get

$$Ly(x) = \sum_{i=0}^{n-2} (c_{i+2} y_{i+2} - a y_i) x^i - a y_{n-1} x^{n-1} - a y_n x^n \tag{2.6}$$

with

$$c_i := i(i-1). \tag{2.7}$$

Having two boundary conditions we must enclose $Ly(x)$ by an interval polynomial of degree $n-2$. We do this with the inclusions

$$x^n \in C x^{n-2}, \quad x^{n-1} \in C x^{n-3} \tag{2.8}$$

with

$$C := [0, 1]. \tag{2.9}$$

Thus we obtain

$$
\begin{aligned}
Ly(x) \in \sum_{i=0}^{n-4} (c_{i+2} y_{i+2} - a y_i) x^i & \\
+ (c_{n-1} y_{n-1} - a y_{n-3} - a C y_{n-1}) x^{n-3} + (c_n y_n - a y_{n-2} - a C y_n) x^{n-2}.
\end{aligned}
\tag{2.10}
$$

The comparison of coefficients in (2.10) and (2.4) yields the $n-1$ equations

$$-aY_i + c_{i+2} Y_{i+2} = G_i, \qquad (i = 0, 1, \ldots, n-4),$$
$$-aY_{n-3} + [c_{n-1} - aC] Y_{n-1} = G_{n-3},$$
$$-aY_{n-2} + [c_n - aC] Y_n = G_{n-2}.$$

(2.11)

The boundary conditions (2.3b) add the two equations

$$Y_0 - Y_1 + Y_2 - + \ldots - Y_{n-1} + Y_n = \alpha,$$
$$Y_0 + Y_1 + Y_2 + \ldots + Y_{n-1} + Y_n = \beta,$$

(2.12)

or equivalently

$$Y_0 + Y_2 + Y_4 + \ldots + Y_{n-2} + Y_n = \frac{\beta + \alpha}{2},$$

$$Y_1 + Y_3 + Y_5 + \ldots + Y_{n-1} = \frac{\beta - \alpha}{2}.$$

(2.13)

The system (2.10), (2.13) of $n+1$ equations can be split into two independent systems containing the equations with even and with odd subscripts resp. Since we are interested in Y_0 only, we consider the system with even subscripts

$$-aY_i + c_{i+2} Y_{i+2} = G_i, \qquad i = 0, 2, 4, \ldots, n-4,$$
$$-aY_{n-2} + [c_n - aC] Y_n = G_{n-2},$$
$$Y_0 + Y_2 + Y_4 + \ldots + Y_n = \frac{\alpha + \beta}{2}.$$

(2.14)

By elimination of $Y_2, \ldots, Y_n$ one gets

$$Y_0 = \frac{\alpha + \beta}{2A} + \frac{B}{A}$$

(2.15)

with

$$A = 1 + \frac{a}{2!} + \frac{a^2}{4!} + \frac{a^3}{6!} + \ldots + \frac{a^{\frac{n}{2}-1}}{(n-2)!} + \frac{a^{\frac{n}{2}}}{n!} \frac{1}{1 - aC/c_n},$$

(2.16)

$$B = -\left\{ \frac{G_0^*}{c_2} + \frac{G_2^*}{c_4} + \frac{G_4^*}{c_6} + \ldots + \frac{G_{n-4}^*}{c_{n-2}} + \frac{G_{n-2}^*}{c_n} \frac{1}{1 - aC/c_n} \right\},$$

(2.17)

$$G_0^* = G_0, \qquad G_{2i}^* = G_{2i} + \frac{a}{c_{2i}} G_{2i-2}^*, \qquad i = 1, 2, \ldots, \frac{n-2}{2}.$$

(2.18)

The expression (2.15) is meaningful only, if the interval A is defined and does not contain zero. By choosing n large enough this can be always achieved, if $a \geq 0$. In the limit we have

$$\lim_{n \to \infty} A = \begin{cases} \cosh(\sqrt{a}), & a \geq 0 \\ 1, & a = 0 \\ \cos(\sqrt{-a}), & a \leq 0 \end{cases}$$

(2.19)

Now we consider the differential equation (2.3a) in an arbitrary interval $X = [x_l, x_r]$ with boundary conditions

$$u(x_l) = \alpha, \quad u(x_r) \overset{!}{=} \beta. \tag{2.20}$$

The linear transformation

$$x := x_m + h\xi, \quad x_m := \frac{x_r + x_l}{2}, \quad h := \frac{x_r - x_l}{2} \tag{2.21}$$

maps X onto $[-1, 1]$ and the midpoint $x = x_m$ of X into the midpoint $\xi = 0$ of $[-1, 1]$. Setting

$$v(\xi) := u(x_m + h\xi) \tag{2.22}$$

the boundary value problem (2.3a), (2.20) is transformed into

$$v''(\xi) - a_h v(\xi) \overset{!}{=} f_h(\xi, \quad \xi \in [-1, 1], \tag{2.23a}$$

$$v(-1) \overset{!}{=} \alpha, \quad v(1) \overset{!}{=} \beta, \tag{2.23b}$$

with

$$a_h := h^2 a, \quad f_h(\xi) =: h^2 g(x_m + h\xi). \tag{2.24}$$

Having computed the value Y_0 for this transformed problem, we have the midpoint inclusion

$$u(x_m) \in Y_0 \tag{2.25}$$

for the original problem.

Remark

In all equations containing intervals all values have to be interpreted as intervals. So e.g. in (2.15) α and β mean $[\alpha, \alpha]$, $[\beta, \beta]$ resp. If these intervals are enlarged, also the interval Y_0 is enlarged, at least, it is not diminished. Therefore, we also get an inclusion, if we use enclosing intervals instead of the exact boundary values.

3. A Kind of Difference Method

Denoting as before the basic interval by $[x_l, x_r]$, we consider an equidistant partition

$$h := \frac{x_r - x_l}{n_0}, \quad x_i = x_l + ih \quad (i = 0, 1, \ldots, n_0) \tag{3.1}$$

with an arbitrary integer $n_0 \geq 2$.

Now, let M be a method, which computes a midpoint inclusion in arbitrary intervals from given values at the endpoints. Then, if an inclusion $u(x_{i-1}) \in Y_{i-1}$, $u(x_{i+1}) \in Y_{i+1}$ of the solution $u(x)$ at two points x_{i-1}, x_{i+1} is given, M yields an inclusion $u(x_i) \in Y_i$. Supposing M, for linear problems, to be a linear method, Y_i depends linearly on Y_{i-1}, Y_{i+1}:

$$Y_i = A_{i0} + A_{i1} Y_{i-1} + A_{i2} Y_{i+1} \tag{3.2}$$

with intervals A_{ik} depending on the data of the differential equation.

Since $Y_0 \ni u(x_l)$, $Y_{n_0} \ni u(x_r)$ are known, (3.2) (for $i=1, 2, ..., n_0-1$) can be considered as a system of linear equations for the unknowns $Y_1, Y_2, ..., Y_{n_0-1}$, given in fixed point form. If, starting with an inclusion vector $Y_i^{[0]} \ni u(x_i)$, (3.2) is iterated, then every iteration vector $(Y_i^{[m]})$ includes $u(x_i)$. If the iteration converges to a vector (Y_i^*), then also $Y_i^* \ni u(x_i)$. Under certain conditions [cf. [1], [3a], [3b]) the iteration converges for every starting vector to the same fixed point $Y^* = (Y_1^*, ..., Y_{n_0-1}^*)^T$; e.g. if the spectral radius of the absolute value of the matrix in the righthand side of (3.2) is <1. Then we have $Y_i^* \ni u(x_i)$ also, if the starting vector is not enclosing; but a stopping criterium for the iteration is hard to find.

Under certain conditions (cf. [3a], Satz 3 or [3b], Theorem 2) the fixed point Y^* is the smallest interval vector enclosing the set $\{x\}$ of all solutions of the point systems contained in the system of interval equations. If that is true for (3.2), then any solution $\bar{Y}$ of the system

$$Y_i - A_{i1} Y_{i-1} - A_{i2} Y_{i+1} = A_{i0}, \quad i=1, 2, ..., n_0-1 \tag{3.3}$$

obtained by a Gauß-algorithm will enclose Y^* and give an inclusion $u(x_i) \in \bar{Y}_i$.

As an example we consider our model problem

$$u''(x) - a u(x) \overset{!}{=} g(x), \quad u(x_l) \overset{!}{=} \alpha, \quad u(x_r) \overset{!}{=} \beta. \tag{3.4}$$

As shown in Section 2 the coefficients of (3.2) are

$$A_{i1} = A_{i2} = \frac{1}{2 A_h}, \quad A_{i0} = \frac{B_i}{A_h}. \tag{3.5}$$

Here A_h and B_i are determined by (2.16)–(2.18), if a is substituted by $h^2 a$ and the G_i by the coefficients of an interval polynomial enclosing $h^2 g(x_i + h\xi)$.

In matrix notation (3.2) is

$$Y = AY + B, \quad B := (B_1, ..., B_{n_0-1})/A_h \tag{3.6}$$

with

$$A = \frac{1}{2 A_h} \begin{bmatrix} 0 & 1 & & & \\ 1 & 0 & 1 & & \\ 0 & 1 & 0 & 1 & \\ & & & \ddots & \\ & & & 1 & 0 \end{bmatrix} \tag{3.7}$$

For $a=0$ we have $A_h=1$ and A is a well known point matrix with $\rho(A) = \rho(|A|) < 1$. For $a>0$ we can choose n_0 or n so large that $|A_h| > 1$ and therefore $\rho(|A|) < 1$. Therefore, for $a \geq 0$ (and n_0 or n large enough) the iteration

$$Y_i^{[m+1]} = \frac{1}{2 A_h} Y_{i-1}^{[m+1]} + \frac{1}{2 A_h} Y_{i+1}^{[m]} + B_i, \quad m = 0, 1, 2, ... \tag{3.8}$$

converges to the fixed point Y^* of (3.6).

For the system

$$A^* Y = B, \qquad A^* := I - A \tag{3.9}$$

all conditions of Satz 3 in [3a] are fulfilled. Therefore $Y^* \in \bar{Y}$ holds for the solution $\bar{Y}$ of (3.9) obtained by Gauß-elimination, which is done very quickly, since A^* is tridiagonal.

4. A Numerical Example. Modifications

Among others the method was tested by a problem which served also in [4] for testing several methods for solving (approximately) boundary value problems. The result there was that most methods give poor results with the exception of a multiple shooting method. We consider a somewhat generalized version of the problem treated in [4]:

$$u''(x) - \kappa^2 u(x) = g(x), \qquad u(0) = u(1) = 0 \tag{4.1}$$

with

$$g(x) := \frac{\kappa^2}{2} + \frac{\lambda^2 + \kappa^2}{2} \cos(\lambda x), \qquad \lambda := 2\pi \tag{4.2}$$

where κ is an arbitrary, real parameter (with $\kappa = 20$ in [4]). The solution is

$$u(x) = \frac{1}{1 + e^{-\kappa}} \{e^{\kappa(x-1)} + e^{-\kappa x}\} - \frac{1}{2} \{\cos(\lambda x) + 1\} \tag{4.3}$$

which can also be written as

$$u(x) = \frac{\cosh(\kappa t)}{\cosh\left(\frac{\kappa}{2}\right)} + \frac{1}{2} \cos(\lambda t) - \frac{1}{2}, \qquad t := x - \frac{1}{2}. \tag{4.4}$$

Using PASCAL-SC the computations were performed on a 13-digit computer SAM 68K from KWS. We give some results for the case $\kappa = 20$ treated in [4]:

n_0	2	2	2	20	20	20
r	10	20	40	4	10	30
W	$5E-5$	$2E-8$	$1E-12$	$1E-4$	$5E-11$	$5E-12$

Here n_0 (as above) denotes the number of subintervals and r the degree of the interval polynomial enclosing $g(x)$ such that $n := r + 2$ is the degree of the Ansatz polynomial in every subinterval. W (rounded to 1 digit) is the maximal width of the enclosing intervals $Y_i \ni u(x_i)$ such that $W/2$ is a bound for the error if the midpoints of the Y_i are used as approximations for the $u(x_i)$.

Sometimes, one is interested only in the value at the midpoint of the interval. Then $n_0 = 2$ can be used and machine accuracy is reached with $r = 40$. If the solution is needed over the whole interval then n_0 has to be chosen larger and r an be taken smaller.

The simplicity of the algorithm for the model problem (3.3) depends on a being a constant. Nevertheless, if a denotes a (continuous) function $a(x)$, one can procede similarly as for constant a by substituting $a(x)$ by an enclosing interval $[a_i]$ in $[x_{i-1}, x_{i+1}]$:

$$a(x) \in [a_i], \quad x \in [x_{i-1}, x_{i+1}] \quad (i = 1, 2, ..., n_0 - 1). \tag{4.5}$$

Then the only difference to the case $a = \text{const}$ is that the elements of the tridiagonal interval matrix are not constant but depend on the row index i. There are now $n_0 - 1$ evaluations instead of 1 needed for the computation of the matrix; that is the only difference in computation time. Obviously, the sharpness of the enclosure of $u(x)$ depends on the sharpness of the enclosures (4.5). This can be increased by taking smaller intervals $[x_{i-1}, x_{i+1}]$, e.g. by increasing n_0. On the other hand, by increasing n_0 the condition of the matrix becomes worse. Therefore, probably, by this method only a limited accuracy can be achieved and we sketch a procedure, which avoids this difficulty and in principle, can be applied to more general problems.

Again we start with the problem (2.3); but now a shall denote a nonconstant polynomial $a(x)$. Using the Ansatz (2.5) $Ly(x)$ now is a polynomial of degree higher than n such that more high powers of x have to be enclosed by interval polynomials of maximal degree $n-2$ and the coefficients of x^k contain more of the unknown y_i. Therefore the matrix of the system of equations resulting from the comparison of coefficients has more non-zero elements and the system cannot be solved for Y_0 by a simple explicit formula like (3.6), (3.7) anymore. Instead, for every interval $[x_{i-1}, x_{i+1}]$ all coefficients $Y_0^{(i)}, Y_1^{(i)}, ..., Y_n^{(i)}$ of the enclosing polynomial $Y^{(i)}(x)$ have to be determined from a system of $n+1$ equations containing $Y_0^{(i-1)}, Y_0^{(i+1)}$ on the "righthand side". Though we are interested only in $Y_0^{(i)}$ ($i = 1, 2, ..., n_0 - 1$), we have to solve a system of $(n_0 - 1)(n+1)$ equations for all $Y_k^{(i)}$. Because of its special structure this system can be solved by some kind of block iteration. Starting with given (possibly enclosing) values $Y_0^{(i)}$ ($i = 1, ..., n_0 - 1$) for every i new values $Y_0^{(i)}, Y_1^{(i)}, ..., Y_n^{(i)}$ are determined from a system of order $i + 1$ by Gauß-elimination.

If $a(x)$ is not a polynomial but an arbitrary (sufficiently smooth) function, then a similar procedure may be applied by substituting $a(x)$ by an enclosing interval polynomial in every interval $[x_{i-1}, x_{i+1}]$.

If the given functions of the differential equation, e.g. $g(x)$ in (3.3), are not smooth enough we cannot hope to increase the sharpness of the enclosure by increasing the degree n of the Ansatz polynomial. Then only decreasing the length of the intervals $[x_{i-1}, x_{i+1}]$, i.e. increasing the partition number n_0 can help. By this not only the computation time is increased, but more important, the condition of the matrix becomes worse. Usually the given functions will be nonsmooth only in the neighbourhood of one or a few isolated points, e.g. of $x = 0$ if the function depends on $|x|$. In such cases it is advisable to use a small stepwidth h only in the neighbourhood of these points and larger values of h in the remaining interval. By a transition from h to h/m or from h/m to h ($m \in \mathbb{N}$) in the model problem the bandwidth of the matrix (3.7) is increased, but not the number of non-zero elements. E.g. if $m = 2$ in some rows the right element will be moved one column to the right and in some other rows the left element one place to the left. The system can be solved by slight modifications of the tridiagonal Gauß-algorithm.

References

[1] Alefeld, G., Herzberger, J.: Introduction to Interval Computations. Academic Press, New York, 1983.
[2] Kaucher, E.W., Miranker, W.L.: Self-Validating Numerics for Function Space Problems. Academic Press, New York, 1984.
[3a] Mayer, G.: Reguläre Zerlegungen und der Satz von Stein und Rosenberg für Intervallmatrizen. Habilitationsschrift. Institut für Angewandte Mathematik der Universität Karlsruhe, 1986.
[3b] Mayer, G.: Enclosing the Solutions of Systems of Linear Equations by Interval Iterative Processes. This volume.
[4] Stoer, J., Bulirsch, R.: Introduction to Numerical Analysis. Springer, New York, Heidelberg, Berlin, 1980.
[5] Weissinger, J.: An Enclosure Method for Differential Equations. In: Kaucher, E., Kulisch, U. and Ullrich, Ch. (Eds): Computerarithmetic. Scientific Computation and Programming Languages. B.G. Teubner, Stuttgart, 1987.

Prof. Dr. Johannes Weissinger
Institut für Angewandte Mathematik
Universität Karlsruhe
Kaiserstrasse 12
D-7500 Karlsruhe 1
Federal Republic of Germany

Computing, Suppl. 6, 33–46 (1988)

A Self-Validating Method for Solving Linear Programming Problems with Interval Input Data

Christian Jansson, Hamburg-Harburg

Abstract — Zusammenfassung

A Self-Validating Method for Solving Linear Programming Problems with Interval Input Data. Linear programming problems are very important in many practical applications. They are usually solved by the simplex method. The computational results are, in general, good approximations to the solution of the problem. However, in some cases the computed approximation may be wrong due to round-off and cancellation errors. In practice it occurs frequently that the input data of a linear programming problem are not known exactly but are afflicted with tolerances. In this case it has to be precisely defined what a "solution" to such a problem is. A sensitivity or postoptimality analysis is necessary.

In the following a method for linear programming problems with interval input data is described which computes guaranteed lower and upper bounds for all optimal vertices and the optimal value. The method controls rigorously all round-off errors and gives an automatic sensitivity analysis. As an example a diet problem is treated to demonstrate how the method in principle works.

Eine Einschließungsmethode zur Lösung linearer Optimierungsprobleme mit Intervalleingabedaten. Lineare Optimierungsprobleme treten sehr häufig in der Praxis auf. Eine effiziente Methode zur Lösung solcher Probleme ist das Simplexverfahren. Die näherungsweise berechneten Lösungen approximieren im allgemeinen die exakte Lösung des Problems gut. Allerdings kann das Auftreten von Rundungs- sowie Auslöschungsfehlern auch zu völlig falschen Resultaten führen.

Bei den meisten praktischen Anwendungen sind die Eingabedaten toleranzbehaftet. In solchen Situationen ist eine Sensitivitätsanalyse oder auch postoptimale Analyse notwendig.

Im folgenden wird ein Algorithmus für lineare Optimierungsprobleme mit Intervalleingabedaten angegeben, der unter sehr allgemeinen Voraussetzungen strenge Einschließungsintervalle für die Komponenten aller auftretenden optimalen Ecken und für den optimalen Zielfunktionswert berechnet. Anhand eines Ernährungsproblems wird gezeigt, wie mit dieser Methode eine automatische Sensitivitätsanalyse mit Rundungsfehlerkontrolle durchgeführt werden kann.

1. Introduction

In economical and physical systems it occurs frequently that the input data are not known precisely, they are uncertain i.e. afflicted with tolerances. The interval mathematics described in several books (Alefeld, Herzberger [2], Kulisch [12, 13], Moore [16]) uses intervals as input data and supplies techniques for producing reliable numerical results of rigorous error bounds for the solution of a given problem.

In the following, R denotes the set of real numbers. An interval is the closed and bounded set of real numbers

$$[a] := [\underline{a}, \bar{a}] := \{t \in \mathbf{R} \mid \underline{a} \le t \le \bar{a}\}$$

where $\underline{a} \le \bar{a}$. If $* \in \{+, -, \cdot, /\}$ is a binary operation on $\mathbf{R}$, then

$$[a] * [b] := \{z \in \mathbf{R} \mid z = a * b,\ a \in [a],\ b \in [b]\} \tag{1}$$

defines a binary operation on the set of intervals (in the real case the power set operations and interval operations coincide). In the case of division it is assumed that $0 \notin [b]$. These operations can easily be realized on a digital computer. An *interval vector* is a tupel $[x] = ([x_i])$ of intervals and an *interval matrix* is a matrix $[A] = ([a_{ij}])$ the coefficients of which are intervals. The operations for interval vectors and interval matrices are defined analogously to the operations for real vectors and matrices; only the real binary operations must be replaced by the interval binary operation (1).

One of the basic problems which arise in practice is to solve a system of linear equations the coefficients of which are not known exactly. This means to solve the system

$$[A] x = [b] \tag{2}$$

where $[A]$ is an $m \times m$ interval matrix and $[b]$ is an interval vector with m components. $[A]$ and $[b]$ define a set of real problems

$$A x = b \tag{3}$$

where $A \in [A]$, $b \in [b]$. To solve these systems of linear equations means in the following to compute an interval vector $[x]$ such that the *solution set* $L := \{x \in \mathbf{R}^n \mid A x = b,\ A \in [A],\ b \in [b]\} \subseteq [x]$. A necessary condition for the existence of such a vector $[x]$ is that all real matrices $A \in [A]$ are nonsingular. $[x]$ is called *inclusion vector* or shortly an *inclusion*. A *self-validating method* for problem (2) is an algorithm producing an interval vector $[x]$ *on a computer* with $L \subseteq [x]$ and with the guarantee that for all real problems (3) exists a unique solutions or otherwise giving a warning that no such interval vector could be found. The latter is the case for example if $[A]$ contains a singular (real) matrix. Recently (Rump [21, 22, 23], ACRITH [1], PASCAL-SC [4]) self-validating methods were developed computing very efficiently and with high accuracy inclusion vectors for systems of linear equations with interval input data and for many other numerical problems. The algorithms for linear systems are the basis of the following self-validating method for linear programming problems.

2. Basic Concepts of Linear Programming

A linear programming problem consists of a linear function to be minimized (or maximized), subject to linear equality and inequality constraints. These problems are very important in practice. For example, Lovasc ([14]) writes: "If one would take statistics about which mathematical problem is using up most of the computer time in the world, then (not counting database handling problems like sorting and searching) the answer would probably be linear programming."

Any linear programme can be put by well known transformations into *standard form*

$$\text{Max}\{c^t x \mid x \in X\}, \quad X := \{x \in \mathbf{R}^n \mid Ax = b, \, x \geq 0\} \tag{4}$$

where $A = (a_1, \ldots, a_n)$ is a real $m \times n$ matrix with column vectors $a_1, \ldots, a_n \in \mathbf{R}^m$, $b \in \mathbf{R}^m$, $c \in \mathbf{R}^n$ and $m < n$. The input data of (4) are given by the tripel $P = (A, b, c) \in \mathbf{R}^{mn+m+n}$. The function $z := c^t x$ is called the *objective function* of (4). X is called the *set of feasible solutions*. X is a convex polyhedron which is the intersection of the affine variety $\{x \in \mathbf{R}^n \mid Ax = b\}$ with the positive orthant. A point $x \in \mathbf{R}^n$ is called *optimal*, if $x \in X$ and $c^t x \geq c^t x'$ for all $x' \in X$. If such a point exists the value $z_{opt} := \text{Max}\{c^t x \mid x \in X\}$ is called the *optimal value*.

A main theorem (cf. [5], [6], [17]) in linear programming tells that there is an optimal solution with at most m components unequal to zero, if there does exist an optimal solution. There is only a finite number of such solutions. The simplex method of Dantzig is an algorithm to compute such an optimal solution efficiently.

Solutions of $Ax = b$ where $n - m$ variables are zero can be described by partitioning $A = (a_1, \ldots, a_n)$ into two submatrices A_B and A_N, where $A_B = (a_{\beta_1}, \ldots, a_{\beta_m})$ is a nonsingular $m \times m$ submatrix with the column vectors a_{β_i} of A, and $A_N = (a_{\gamma_1}, \ldots, a_{\gamma_{n-m}})$ is the $m \times (n-m)$ submatrix with the remaining column vectors of A. The set of indices $B = \{\beta_1, \ldots, \beta_m\} \subseteq \{1, \ldots, n\}$ which determines the nonsingular matrix A_B is called a *basic-index-set* of (4) and $N := \{\gamma_1, \ldots, \gamma_{n-m}\} := \{1, \ldots, n\} \setminus B$ is the set of *nonbasic indices*. B and N describe the partitioning of the matrix A.

For a given partitioning B, N the linear equation $Ax = b$ can be expressed in the form

$$A_B x_B + A_N x_N = b \tag{5}$$

where $x = (x_B, x_N)^t$ is partitioned analogously. By multiplying equation (5) with the inverse A_B^{-1} we get

$$x_B = A_B^{-1} b - S_N x_N, \quad S_N := A_B^{-1} A_N. \tag{6}$$

Partitioning $c = (c_B, c_N)^t$ similarily, the objective function can then be expressed in the form

$$c^t x = c_B^t x_B + c_N^t x_N = c_B^t (A_B^{-1} b - S_N x_N) + c_n^t x_N$$
$$= c_B^t A_B^{-1} b - (c_B^t S_N - c_N^t) x_N$$

and therefore

$$c^t x = c_B^t A_B^{-1} b - d_N^t x_N, \quad d_N := S_N^t c_B - c_N = A_N^t (A_B^t)^{-1} c_B - c_N. \tag{7}$$

Thus both the m variables x_B and the value of the objective function are determined by the $n - m$ variables x_N.

The solution, where the $n - m$ variables x_N are zero

$$x(B) := (x_B, x_N)^t, \quad x_B := A_B^{-1} b, \quad x_N := 0 \tag{8}$$

is called a *basic solution* (corresponding to the basic-index-set B). If in addition $x(B) \geq 0$ (this is the case if $x_B := A_B^{-1} b \geq 0$) then $x(B) \in X$ and is called a *basic feasible solution*. The value of the objective function is $z(B) := c^t x(B) = c_B^t A_B^{-1} b$. Geometrically the basic feasible solutions correspond to the vertices of the convex polyhedron X. The vector d_N in (7) is called the vector of *reduced costs*, since it indicates how the objective function depends on x_N.

If for a given partitioning B, N the vector $x(B)$ is a basic feasible solution and $d_N \geq 0$ then $x(B)$ is optimal: because of (7) for any $x' \in X$ the value

$$c^t x' = c_B^t A_B^{-1} b - d_N^t x_N' \leq c_B^t A_B^{-1} b = c^t x(B)$$

since $x_N' \geq 0$ and $d_N \geq 0$. Analogously it is easy to see that $x(B)$ is the unique optimal solution of (4) if $d_N > 0$ and $x_B := A_B^{-1} b > 0$.

A basic-index-set B is called *optimal*, if $x(B)$ is a basic feasible solution and $d_N \geq 0$. V_{opt} designates the set of optimal basic-index-sets and $X_{opt} := \{ x(B) \mid B \in V_{opt} \}$ is the set of optimal basic feasible solutions of (4). Geometrically the elements of X_{opt} correspond to the optimal vertices of the convex polyhedron X and any optimal solution of (4) is a convex combination of the optimal basic feasible solutions, if the set of optimal solutions is bounded. Therefore the interesting output data of the linear programming problem (4) are: *the optimal value z_{opt}, the set of optimal basic-index-sets V_{opt} and the set of optimal basic feasible solutions X_{opt}*.

The simplex method determines efficiently the optimal value z_{opt} and one optimal basic-index-set B with the corresponding optimal basic feasible solution $x(B)$. This is no disadvantage because in most cases $d_N > 0$ and $x_B := A_B^{-1} b > 0$ for real input data, and therefore there exists only one optimal basic-index-set.

However, in unfavourable cases the computed approximation may be drastically wrong due to round-off errors during computation and uncertainties of the input data. It is easy to give examples (cf. Krawzcyk [11]) where the simplex algorithm computes an approximation of an optimal solution although the set of feasible solutions X is empty or the objective function is unbounded on X. Therefore a sensitivity or postoptimality analysis and error analysis is an important part for linear programming problems. For example, Murtagh ([17], page 62) writes: "It is often said that postoptimality analysis is the most important part of LP calculations, and it is not hard to see why this conclusion is reached. Most of the coefficients which appear in an LP problem are not known exactly, and in practice are usually best estimates of the value that the coefficients should be."

3. A Self-Validating Method for Linear Programming Problems

To perform sensitivity and error analysis of a linear programming problem the input data are described by the tripel

$$[P] = ([A], [b], [c]) \tag{9}$$

where $[A]$ is an $m \times n$ interval matrix, $[b]$ is an interval vector with m components and $[c]$ is an interval vector with n components. The tripel $[P]$ defines a set of linear programming problems with real input data $P \in [P]$. This set can be viewed as a linear parametric optimization problem where each coefficient varies independently between the given lower and upper interval bound. In the following we write $z_{opt}(P)$, $V_{opt}(P)$, $X_{opt}(P)$, $X(P)$, ... to indicate the dependence on $P \in [P]$. Obviously, from the geometrical and algebraic point of view, there will, in general, exist problems with $P \in [P]$ which have different optimal basic feasible solutions. In the following we treat the problem to compute inclusions for $z_{opt}(P)$, $V_{opt}(P)$, $X_{opt}(P)$ for the set of all linear programming problems $P \in [P]$ defined by the tripel $[P] = ([A], [b], [c])$. That means to calculate an interval $[z]$, a set of index-sets $Z = \{B^1, ..., B^s\}$ and interval vectors $[x^1], ..., [x^s]$ such that the following three conditions are satisfied:

1° For all $P \in [P]$: $z_{opt}(P) \in [z]$
2° For all $P \in [P]$: $B \in V_{opt}(P) \Rightarrow B \in Z$
3° For all $P \in [P]$: $x \in X_{opt}(P) \Rightarrow x \in [x^i]$ where $i \in \{1, ..., s\}$

The following self-validating method either computes such inclusions $[z]$, $[x^1], ..., [x^s]$ and Z where conditions 1°, 2°, 3° are fulfilled and guarantees that for all real problems $P \in [P]$ there exist optimal solutions, or gives a warning that no inclusion could be calculated. The latter is the case, for example, if there is a linear programming problem with $P \in [P]$ or nearby such that the set of feasible solutions $X(P)$ is empty or the objective function is unbounded on $X(P)$. The algorithm delivering solutions with the guarantees mentioned above is efficiently executable on digital computers.

We allot to $[P]$ the graph $G_{opt}([P]) := (V_{opt}([P]), E_{opt}([P]))$ with the node set

$$V_{opt}([P]) := U \{V_{opt}(P) \mid P \in [P]\} \tag{10}$$

and the edge set

$$E_{opt}([P]) := \left\{ \{B, B'\} \,\middle|\, \begin{matrix} B, B' \text{ differ in exactly} \\ \text{one index and there exists a} \\ P \in [P] \text{ with } B, B' \in V_{opt}(P) \end{matrix} \right\} \tag{11}$$

$G_{opt}([P])$ is called the *representation graph* of $[P]$. Two nodes B^1, B^2 are said to be *neighboured* if $\{B^1, B^2\} \in E_{opt}([P])$. For $B \in V_{opt}([P])$ the set $N(B)$ designates the set of all nodes B' in $V_{opt}([P])$ which are neighboured to B. The method to be presented can be viewed as a graph-search-method (cf. Papadimitriou, Steiglitz [19]) applied to the graph $G_{opt}([P])$. In the first step a starting node $B_{start} \in V_{opt}([P])$ is computed. Then all nodes in $N(B_{start})$ are computed. In the following all nodes in $N(B')$ are computed where $B' \in N(B_{start})$, and so forth. Obviously, all nodes in $V_{opt}([P])$ are calculated if the graph $G_{opt}([P])$ is connected. To compute the sets $N(B)$ we introduce interval tableaus corresponding to basic-index-sets of $[P]$.

An index-set $B = \{\beta_1, ..., \beta_m\}$ is called a *basic-index-set* of $[P] = ([A], [b], [c])$ if all real matrices $A_B \in [A_B] := ([a_{\beta_1}], ..., [a_{\beta_m}])$ are nonsingular. $[a_{\beta_1}], ..., [a_{\beta_m}]$ are the columns of the interval matrix $[A]$ which corresponds to the indices $\beta_1, ..., \beta_m$.

Let $B=\{\beta_1, \ldots, \beta_m\}\subseteq\{1, \ldots, n\}$ be an index-set and $[x_B]$, $[y(B)]$, $[s_\gamma]$ with $\gamma\in N:=\{\gamma_1, \ldots, \gamma_{n-m}\}:=\{1, \ldots, n\}\setminus B$ be inclusion vectors for the solution sets of the systems of interval linear equations

$$[A_B]\,x_B=[b]$$
$$[A_B]^t\,y(B)=[c_B] \tag{12}$$
$$[A_B]\,s_\gamma=[a_\gamma]$$

which are computed by a self-validating method. Then it is verified that all real matrices $A_B\in[A_B]$ are nonsingular. We define

$$[d_N]:=[A_N]^t[y(B)]-[c_N] \tag{13}$$
$$[x(B)]:=([x_B], x_N)^t, \qquad x_N:=0 \tag{14}$$
$$[S_N]:=([s_{\gamma_1}], \ldots, [s_{\gamma_{n-m}}]) \tag{15}$$
$$[z(B)]:=[c_B]^t[x_B]\cap[b]^t[y(B)]. \tag{16}$$

Obviously, the interval vectors $[d_N]$, $[x(B)]$, the interval matrix $[S_N]$ and the interval $[z(B)]$ are inclusions for the corresponding vectors d_N, $x(B)$ and the matrix S_N defined in (6), (7), (8) and the value of the objective function. We comprise these intervals in the form

$$[T(B)]:=\left(\frac{[z(B)] \mid [d_N]^t}{[x_B] \mid [S_N]}\right)=\begin{pmatrix} [z(B)] & \mid & [d_{\gamma_1}] & \cdots & [d_{\gamma_{n-m}}] \\ \hline [x_{\beta_1}] & \mid & [s_{\beta_1\gamma_1}] & \cdots & [s_{\beta_1\gamma_{n-m}}] \\ \vdots & \mid & \vdots & & \vdots \\ [x_{\beta_m}] & \mid & [s_{\beta_m\gamma_1}] & \cdots & [s_{\beta_m\gamma_{n-m}}] \end{pmatrix} \tag{17}$$

and call $[T(B)]$ the *interval tableau* corresponding to the basic-index set B.

Theorem 1: Let B be a basic-index-set of $[P]$ and $[T(B)]$ the corresponding interval tableau.

1. If the conditions

$$[x_B]=[\underline{x}_B, \bar{x}_B]>0 \quad \text{and} \quad [d_N]=[\underline{d}_N, \bar{d}_N]>0 \tag{18}$$

are satisfied then for all linear programming problems with $P\in[P]$ there exists a unique optimal solution which is contained in $[x(B)]$. Moreover $V_{\text{opt}}([P])=\{B\}$ and $z_{\text{opt}}(P)\in[z(B)]$.
2. If $B\in V_{\text{opt}}([P])$ then

$$\bar{x}_B\geq0 \quad \text{and} \quad \bar{d}_N\geq0 \tag{19}$$

Proof:
1) If $[x_B]>0$ and $[d_N]>0$ then $x_B:=A_B^{-1}b>0$ and $d_N:=A_N^t(A_B^t)^{-1}c_B-c_N>0$ for each linear programming problem with $P\in[P]$ because of the inclusion monotonicity of interval operations. Therefore the optimal basic-index-set B is unique (compare section 2.) for all problems with $P\in[P]$. Therefore $V_{\text{opt}}([P])=\{B\}$, and from the definitions follows directly that $[x(B)]$ contains all optimal solutions and $[z(B)]$ contains the optimal values for all $P\in[P]$.

2) If $B \in V_{\mathrm{opt}}([P])$, then there exists a $P = (A, b, c) \in [P]$ with $B \in V_{\mathrm{opt}}(P)$. From $B \in V_{\mathrm{opt}}(P)$ it follows that $x_B := A_B^{-1} b \geq 0$ and $d_N \geq 0$. Since $x_B \in [x_B]$ and $d_N \in [d_N]$ it follows $\bar{x}_B \geq 0$ and $\bar{d}_N \geq 0$. $\quad\square$

The first-part of Theorem 1 was proved by Krawzcyk [11]. If the conditions $[x_B] > 0$, $[d_N] > 0$ are satisfied, then $[P]$ is called *basisstable*. The second part of Theorem 1 gives a necessary condition that B is an optimal basic-index-set for a problem $P \in [P]$. Such basic-index-sets are only interesting if $[P]$ is not basisstable. The following theorem shows how to compute an inclusion for $N(B)$ where $B \in V_{\mathrm{opt}}([P])$.

Theorem 2: Let B be a basic-index-set of $[P]$ with $\bar{x}_B \geq 0$, $\bar{d}_N \geq 0$ and $[T(B)]$ the corresponding interval tableau. Let $\tilde{N}(B)$ be defined as the set of all index-sets $B' = (B \setminus \{\beta\}) \cup \{\gamma\}$ with $\beta \in B$, $\gamma \in N$ which satisfies one of the following conditions:

$$o \in [d_\gamma], \bar{s}_{\beta\gamma} > 0 \quad \text{and} \quad \frac{x_\beta}{\bar{s}_{\beta\gamma}} \leq \mathrm{Min}\left\{ \frac{\bar{x}_{\beta'}}{\underline{s}_{\beta'\gamma}} \,\middle|\, \underline{s}_{\beta'\gamma} > 0, \beta' \in B \right\} \tag{20}$$

$$o \in [x_\beta], \underline{s}_{\beta\gamma} < 0 \quad \text{and} \quad \frac{d_\gamma}{\underline{s}_{\beta\gamma}} \geq \mathrm{Max}\left\{ \frac{\bar{d}_{\gamma'}}{\bar{s}_{\beta\gamma'}} \,\middle|\, \bar{s}_{\beta\gamma'} < 0, \gamma' \in N \right\} \tag{21}$$

If $B \in V_{\mathrm{opt}}([P])$, then $\tilde{N}(B) \supseteq N(B)$.

With Theorem 2 we can describe the following self-validating method for linear programming problems with interval input data $[P]$:

(1): compute $B_{\mathrm{start}} \in V_{\mathrm{opt}}([P])$;
$\quad\quad S := \emptyset$, $Z := \emptyset$, $B := B_{\mathrm{start}}$;
$\quad\quad$ goto (4);

(2): if $S = \emptyset$ then STOP

(3): choose $B \in S$; $S := S \setminus \{B\}$

(4): a) compute $[T(B)]$;
$\quad\quad\quad$ if $\{[T(B)]$ is not computable or either one of
$\quad\quad\quad\quad$ the following conditions
$\quad\quad\quad\quad\quad$ (i) $\;0 \in [x_\beta]$ with $\beta \in B \Rightarrow$ there exists a
$\quad\quad\quad\quad\quad\quad \gamma \in N$ with $[s_{\beta\gamma}] < 0$
$\quad\quad\quad\quad\quad$ (ii) $\;0 \in [d_\gamma]$ with $\gamma \in N \Rightarrow$ there exists a
$\quad\quad\quad\quad\quad\quad \beta \in B$ with $[s_{\beta\gamma}] > 0$
$\quad\quad\quad\quad$ is not true$\}$
$\quad\quad\quad$ then STOP (WARNING);
$\quad\quad$ b) if $\{$there exists a $\beta \in B$ with $[x_\beta] < 0$ or there
$\quad\quad\quad\quad$ exists a $\gamma \in N$ with $[d_\gamma] < 0\}$
$\quad\quad\quad$ then goto (2);
$\quad\quad$ c) compute $\tilde{N}(B)$ using Theorem 2;
$\quad\quad\quad Z := Z \cup \{B\}$; $S := S \cup \{N(B) \setminus Z\}$;
$\quad\quad\quad$ goto (2);

Theorem 3: If the algorithm described above stops with step (2) then

1. For all linear programming problems with input data $P \in [P]$ there exists an optimal solution and the set of all optimal solutions is bounded. Moreover the optimal value function $z_{opt} : [P] \to \mathbf{R}$ with $z_{opt}(P) := \text{Max}\{c(P)^t x \mid x \in X(P)\}$ is continuous and the graph $G_{opt}([P])$ is connected.
2. The computed set of basic-index-sets $Z = \{B_1, \ldots, B_s\}$, the interval vectors $[x(B_1)], \ldots, [x(B_s)]$ and the interval $[z] := [z(B_1)] \cup \ldots \cup [z(B_s)]$ satisfy 1°, 2°, 3°.

A thorough proof of theorem 2 and theorem 3 would be far too lengthy to be included in these notes (cf. Jansson [9]).

Remarks: 1. A basic-index-set $B_{start} \in V_{opt}([P])$ in step (1) of the algorithm can, for example, be determined in the following way:

Choose some $P = (A, b, c) \in [P]$ and compute an approximation of the optimal solution and the corresponding basic-index-set B for this linear programming problem using the simplex algorithm. Calculate inclusion vectors $[x_B]$, $[y_B]$ for the systems of linear equations

$$A_B x_B = b, \quad A_B^t y_B = c_B \tag{22}$$

If $[x_B] > 0$ and $[d_N] := A_N^t [y_B] - c_N > 0$ then by Theorem 1 B is the unique optimal basic-index-set of the real problem $P = (A, b, c)$. Therefore $B_{start} := B \in V_{opt}([P])$.

2. The algorithm in the presented form stops in step (4) and gives a warning when either $[T(B)]$ is not computable or the conditions (i), (ii) are not true. If $[T(B)]$ is not computable then necessarily the self-validating method for solving the systems (12) gives a warning. This warning indicates that a matrix $A_B \in [A_B]$ is singular or nearby singularity. If condition (i) (resp. (ii)) is not true the warning indicates, for example, that there is a real problem $P \in [P]$ such that $X(P)$ is empty (resp. the objective function is unbounded). In this case it is not possible to compute inclusions for $[P]$.

3. There are many possibilities to choose some $B \in S$ in step (3) of the algorithm. This choice may, in special cases, influence the efficiency of the algorithm. Two extreme strategies are breadth-first-search and depth-first-search. This means that in step (3) the first or last basic-index-set of the list S is selected, respectively.

4. The main work of the algorithm is to compute the interval tableau $[T(B)]$. The only purpose of the interval tableaus is to compute an inclusion of the set of basic-index-sets $N(B)$ by Theorem 2. Computing $N(B)$ it is obviously enough to get inclusion vectors $[x_B]$, $[d_N]$, of the columns of $[S_N]$ with $0 \in [d_\gamma]$ and the rows of $[S_N]$ with $0 \in [x_\beta]$. Since in practical applications $[d_\gamma] > 0$ and $[x_\beta] > 0$ for most indices $\gamma \in N$, $\beta \in B$ it is much less effort to compute only the corresponding columns and rows of $[S_N]$.

5. If the condition in (4) b) of the method is satisfied, then (19) is not true and $B \in V_{opt}([P])$. Therefore B is not admitted to Z.

6. The dual problem of (4) has the form

$$\text{Min}\{b^t y | y \in Y\}, \qquad Y := \{y \in \mathbf{R}^m | A^t y \geq c\} \tag{23}$$

with the same input data $P = (A, b, c)$. It should be mentioned that the algorithm determines automatically inclusions of the optimal vertices of all dual problems with $P \in [P]$. That means for all $P \in [P]$, the following is true: if y is an optimal vertex of the dual problem with $P \in [P]$, then $y \in [y(B^i)]$ where $B^i \in Z$. Notice that $[y(B^i)]$ is the inclusion vector for the corresponding system in (12).

7. In this context, especially for the basisstable case and related questions, the works of Beeck [3], Gal [8], Klatte [10], Machost [15], Rohn [20] and Steuer [24] should be mentioned.

4. The Diet Problem

One of the first applications of linear programming was to determine a daily good nutrition with minimal costs (Dantzig [6], page 625). In the following we examine such a problem with 14 foods and 8 different nutrients respectively vitamines.

Table 1:
Vitamines, nutrients and costs of foods per 100 g (DM = Deutsche Mark, g = gramme)

	Vit. A μg 100 g	Vit. C mg 100 g	Sodium mg 100 g	Potash mg 100 g	Calcium mg 100 g	Carbonhyd. g 100 g	Protein g 100 g	Calories kcal 100 g	Costs DM 100 g
Milk	13	2	47	155	118	5	3	45	0.10
Yoghurt	14	2	45	149	114	6	5	48	0.24
Cheese	150	0	800	95	800	4	26	265	1.25
Roast beaf	0	0	51	340	3	0	19	116	2.10
Pork cutlet	0	0	56	373	2	0	21	156	1.45
Veal cutlet	0	1	83	355	15	0	21	99	1.90
Chicken	0	0	66	264	14	0	23	99	0.70
Cauliflower	21	69	16	311	22	4	3	27	0.20
Carrots	1100	8	45	341	37	7	1	41	0.20
Cabbage	150	10	10	194	23	2	1	17	0.20
Noodles	60	0	7	155	20	72	13	367	0.36
Rice	0	0	6	103	6	79	7	318	0.29
Whole bread	80	7	424	291	56	46	7	213	0.30
Apple purée	7	2	2	96	4	19	0	61	0.14

Table 2:
The daily requirements of nutrients and vitamines chosen for a woman of the age between 19 *and* 35 *years.*

Vit. A	Vit. C	Sodium	Potash	Calcium	Carbonhyd.	Protein	Calories
800 μg	75 mg	2500 mg	3500 mg	800 mg	274 mg	82	2200 kcal

The coefficients of table 1 and table 2 are taken from corresponding tables in [7, 18]. The costs are realistic for stores in Germany in the year 1986.

The linear programming formulation of the above problem is straightforward. The decision variables $x_1, \ldots, x_{14}$ describe the daily consumption of the different foods, a_{ij} ($i=1, \ldots, 8, j=1, \ldots, 15$) is the content of the i-th nutrient (resp. vitamine) in food j (compare table 1), b_i is the daily requirement (compare table 2) and c_j describes the costs of the foods. Obviously, the daily requirements for our target people (women of the age between 19 and 35) are satisfied if the constraints

$$a_{i1}x_1 + \ldots + a_{i14}x_{14} \geq b_i, \quad i=1, \ldots, 8 \tag{24}$$

are fulfilled. The costs are

$$z = c_1 x_1 + \ldots + c_{14} x_{14} \tag{25}$$

and the decision variables x_j are nonnegative. In addition we introduce upper bounds

$$x_j \leq u_j, \quad j=1, \ldots, 14 \tag{26}$$

for decision variables. For milk (resp. yoghurt) the upper bound is 400 g (resp. 300 g) and otherwise 200 g. Moreover the constraint

$$x_4 + x_5 + x_6 + x_7 \geq 100 \tag{27}$$

is added. This constraint means that at least 100 g meat will be consumed.

This problem can be transformed with slack variables into the standard form (4) with $m=23$ and $n=46$. The (rounded) optimal solution computed by the simplex algorithm is: 400 g milk, 156 g cheese, 100 g chicken, 200 g cauliflower, 200 g carrots, 108 g cabbage, 200 g noodles, 61 g rice, 200 g whole bread. The costs are 5.56 DM. The reader may find this solution unsatisfactory on several counts. One possibility to get better solutions is to add new constraints and solve the resulting linear programming problem. If the solution is still unsatisfactory the whole process may be repeated, possibly several times, until a satisfactory menu is found (cf. Chvatal [5], page 184). To proceed in this way takes a lot of computing time and is, in general, far away from optimality because the added constraints do not consider the objective function and cut possibly important parts from the original polyhedron X.

But in view of the unreliability of the data in table 1, 2 it is possible to take intervals as input data and combine the computed "optimal solutions" to a satisfactory menu. For the above example the costs c_j were replaced by the intervals $[c_j] = [c_j - (5/100) c_j, c_j + (5/100) c_j]$. The diameter of $[c_j]$ is 10% of the midpoint of the interval. This diameter is consistent with the variations of the costs in reality. Analogously the data from table 1 and table 2 were replaced by intervals having a diameter of 1% with respect to the midpoint. I am grateful to Mrs. Uschi Maichle who has implemented the algorithm described in part 3 on a CADMUS-PC using the programming language PASCAL-SC(cf. [4]) and computed the diet problem with the above interval input data. The algorithm calculated 8 interval vectors which are listed in table 3 and satisfy the condition 3°.

Table 3:

The calculated 8 interval vectors rounded to 3 figures. Every interval with diameter less than 10^{-11} is replaced by its midpoint.

	$[x^1]$	$[x^2]$	$[x^3]$	$[x^4]$	$[x^5]$	$[x^6]$	$[x^7]$	$[x^8]$
Milk	400	400	400	400	400	400	400	400
Yoghurt	0	0	0	[46, 115]	[6, 70]	[155, 224]	0	[121, 185]
Cheese	[153, 159]	[153, 159]	[153, 159]	[148, 157]	[151, 159]	[142, 159]	[154, 160]	[144, 153]
Roast beef	0	0	0	0	0	0	0	0
Pork cultet	0	0	0	0	0	0	0	0
Veal cutlet	0	0	0	0	0	0	0	0
Chicken	100	100	100	100	100	100	100	100
Cauliflower	200	200	200	200	200	200	200	200
Carrots	200	200	200	200	200	200	200	200
Cabbage	[86, 130]	[107, 155]	[32, 80]	0	0	0	[5, 49]	0
Noodles	200	[70, 87]	[39, 57]	[32, 54]	200	[55, 77]	200	200
Rice	[51, 71]	200	200	200	[11, 36]	200	[16, 36]	[37, 62]
Whole bread	200	200	200	200	200	200	200	200
Apple purée	0	0	200	200	200	0	200	0

These interval vectors represent 8 different menus with 8 different basic-index-sets. These menus can easily be discribed by the tree in Table 4.

Table 4:

The corresponding tree of the 8 interval vectors.

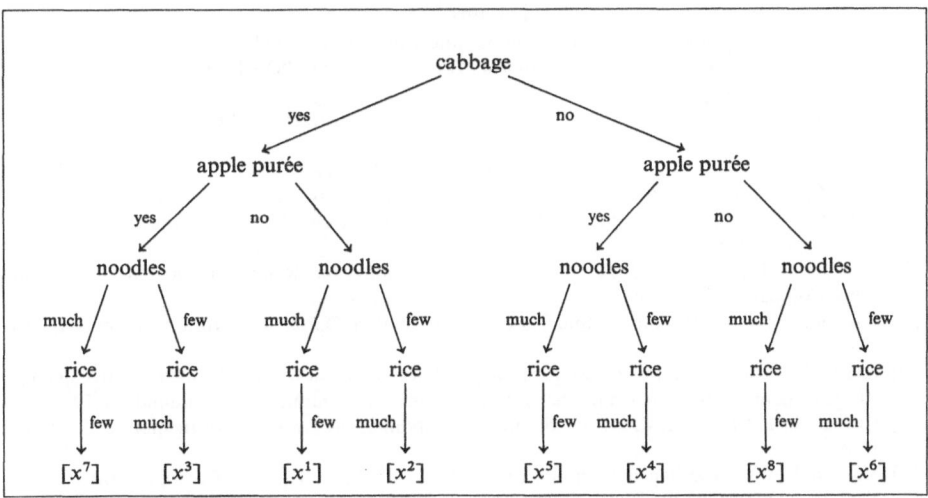

The computed interval for the objective function is $[z] = [5.14, 6.10]$ and it can also be shown that $\{z_{\mathrm{opt}}(P) \mid P \in [P]\}$ contains the interval $[5.21, 5.96]$. The corresponding basic-index-sets are not listed here because they depend on the description of the diet problem in standard form.

The intervals computed by the algorithm give realistic bounds on the behaviour of the diet problem when the input data are not known exactly. For example, a diameter of 10% in the coefficients of the objective function and 1% in the other input data lead to an interval $[z]$ with diameter 17% relative to the midpoint.

For a satisfactory menu one can choose some real representatives of the 8 interval vectors. If the midpoints of the 8 interval vectors are chosen then the original objective function with the real coefficients from table 1 varies between 5.56 DM and 5.62 DM. Obviously, the costs of convex combinations of such vectors do also vary between 5.56 DM and 5.62 DM.

A common technique to perform a sensitivity analysis is to use a Monte Carlo approach. We used 1000 randomly generated diet problems with uniformly distributed input data in the intervals defined above. Only 4 different basic-index-sets were computed, although it can be shown, that $V_{opt}([P])$ has at least 6 different basic-index-sets for the diet problem.

Acknowledgement

I wish to thank Prof. Dr. H. Brakhage, who initiated this research, for many valuable and helpful discussions.

References

[1] ACRITH: IBM High-Accuracy Arithmetic Subroutine Library, Program Description and User's Guide, IBM Program Number 5664−185, 1986.
[2] Alefeld, G., Herzberger, J.: Einführung in die Intervallrechnung, B.I. Wissenschaftsverlag, 1974.
[3] Beeck, H.: Linear Programming with Inexact Data, Bericht Nr. 7830 der Abteilung Mathematik TU München, 1978.
[4] Bohlender, G., Rall, L.B., Ullrich, Ch., Wolff v. Gudenberg, J.: PASCAL-SC, Bibliographisches Institut Mannheim/Wien/Zürich 1986.
[5] Chvatal, V.: Linear Programming, W.H. Freeman and Company, 1983.
[6] Dantzig, G.B.: Lineare Programmierung und Erweiterung, Springer-Verlag, 1966.
[7] Elmadjy, I., Fitsche, D., Cremer, H.-J.: Die große Vitamin und Mineralstofftabelle, Gräfe- und Unzer-Verlag, 1985.
[8] Gal, T.: Postoptimal Analysis, Parametric Programming and Related Topics, Mc Graw-Hill-Book Company, 1979.
[9] Jansson, C.: Zur Linearen Optimierung mit unscharfen Daten, Dissertation, Kaiserslautern, 1985.
[10] Klatte, D.: Lineare Optimierungsprobleme mit Parametern in der Koeffizientenmatrix der Restriktionen, in: Anwendungen der linearen parametrischen Optimierung, Lammatsch, 1979.
[11] Krawczyk, R.: Fehlerabschätzung bei linearer Optimierung, Interval Mathematics, 215−227, 1975.
[12] Kulisch, U.W.: Grundlagen des numerischen Rechnens, B.I. Wissenschaftsverlag, 1976.
[13] Kulisch, U.W., Miranker, W.L.: A New Approach to Scientific Computation, Academic Press New York 1983.
[14] Lovácz, L.: A New Linear Programming Algorithm − Better or Worse Than the Simplex Method?, The Mathematical Intelligencer, Vol. 2, 141−146, 1980.
[15] Machost, B.: Numerische Behandlung des Simplexverfahrens mit intervallmathematischen Methoden, Berichte der GMD Bonn, Nr. 30, 1970.
[16] Moore, R.E.: Methods and Applications of Interval Analysis, SIAM Philadelphia, 1979.
[17] Murtagh, B.A.: Advanced Linear Programming; Computation and Practice, Mc Graw-Hill-Book Company, 1981.
[18] Nährwertbroschüre, Deutsche Lebensmittelwerke GmbH, 1976.

[19] Papadimitrion, C.H., Steiglitz, K.: Combinational Optimization; Algorithms and Complexity, Prentice-Hall, 1982.
[20] Rohn, J.: Solving Interval Linear Systems, Freiburger Intervallberichte 8417, 1−14, 1984.
[21] Rump, S.M.: Kleine Fehlerschranken bei Matrixproblemen, Dissertation, Karlsruhe, 1980.
[22] Rump, S.M.: Solving Algebraic Problems with High Accuracy, Habilitationsschrift, in [13], 1983.
[23] Rump, S.M.: Solving of Linear and Nonlinear Algebraic Problems with Sharp Guranteed Bounds, Computing Suppl. 5, 147−168, 1984.
[24] Steuer, R.E.: Algorithms for Linear Programming Problems with Interval Objective Function Coefficients, Mathematics of Oper. Res., Vol. 6, 333−348, 1981.

Dr. C. Jansson
Informatik III
Technische Universität Hamburg
Eissendorfer Strasse 40
D-2100 Hamburg 90
Federal Republic of Germany

Computing, Suppl. 6, 47–58 (1988)

Enclosing the Solutions of Systems of Linear Equations by Interval Iterative Processes

Günter Mayer, Karlsruhe

Summary — Zusammenfassung

Enclosing the Solutions of Systems of Linear Equations by Interval Iterative Processes. We present a class of iterative processes to enclose the solutions of systems of linear equations $Ax=b$, where the coefficients of A and b are allowed to vary within given intervals. The methods are based on so-called M-splittings and contain many standard iterative processes. We derive results concerning the feasibility, the global convergence, the quality of enclosure and the speed of convergence.

Zur Lösungseinschließung bei linearen Gleichungssystemen durch Intervalliterationsverfahren. Wir behandeln eine Klasse von Iterationsverfahren, die der Lösungseinschließung bei linearen Gleichungssystemen $Ax=b$ dienen. Dabei variieren die Koeffizienten von A und b in gegebenen Intervallen. Die vorgestellten Verfahren basieren auf sogenannten M-Zerlegungen und enthalten viele Standarditerationsverfahren als Spezialfälle. Wir untersuchen die Durchführbarkeit, globale Konvergenz, Einschließungsgüte und Konvergenzgeschwindigkeit dieser Verfahren.

1. Introduction

In classical numerical analysis the solution of linear equations

$$Ax=b \quad (A\in\mathbb{R}^{n\times n}, b\in\mathbb{R}^n) \tag{1}$$

plays a fundamental role. Such equations may arise by the problem itself (finite linear model) or by the numerical method of approximating its solution (linearization, discretization, interpolation etc.). Especially the discretization of linear initial and boundary value problems leads to equations like (1) with a large nonsingular matrix A which has many zero entries. In this case solving (1) by direct methods such as Gaussian algorithm can be very time consuming. Therefore one often replaces them by iterative methods. A large class of such methods is based on an appropriate splitting $A=M-N$ of the matrix A assuming M to be nonsingular. Equation (1) is then equivalent to $Mx=Nx+b$, inducing the iterative process

$$Mx^{k+1}=Nx^k+b, \quad k=0,1,\dots. \tag{2}$$

Examples are Jacobi method, Gauß-Seidel method, relaxation methods, ADI method etc. Equation (2) is, of course, of little use, if it is as difficult to solve as (1). Therefore one often chooses triangular matrices M which involve a straightforward strategy to obtain the new iterate, or one chooses band matrices M with a small band to guarantee storage savings when applying the Gaussian algorithm to equation (2).

It is well known (cf. [20]) that method (2) is globally convergent to the solution of (1) if and only if the spectral radius $\rho(M^{-1}N)$ of the iteration matrix $M^{-1}N$ is less than one, and that $\rho(M^{-1}N)$ is a measure for the speed of convergence (see [20] e.g.). But $\rho(M^{-1}N)$ is not easily available! Therefore results concerning "a good choice" of M are difficult to derive.

Fortunately the matrices A appearing in practice often show some properties which allow a better access to $\rho(M^{-1}N)$. Thus the discretization of elliptic boundary value problems normally leads to M-matrices (cf. [8]) to be defined in Section 2. For such matrices one can easily find splittings $M-N$ with M being an M-matrix and with all entries of N being nonnegative. Jacobi method and Gauß-Seidel method are based on such splittings for which the term M-splitting was introduced by Schneider in [18]. Since M-splittings are so-called regular splittings, it is well known (cf. [20]) that (2) is globally convergent and that $\rho(M^{-1}N)$ depends monotonically on the entries of M and N, allowing comparison results for different splittings. But even in this case one question remains unanswered: How close does the numerical solution approach the exact one?

In classical numerical analysis one usually measures the quality of approximation by the defect Ax^k-b which is not always a reliable quantity. Some authors like Collatz [6], Albrecht [1] and Schröder [19] decompose A into two matrices with which they build up two sequences of iterates converging monotonically to the solution x of (1) while bounding it from above and below, respectively. Alefeld showed in [2] that this method is equivalent to an iterative process using interval arithmetic. Thus if an enclosure of x is wanted or if some quality of approximation should be guaranteed the use of interval analysis is appropriate combined with an accurate arithmetic as described in [9] e.g. It exposes a powerful tool to minimize and to enclose the inevitable rounding errors and to derive bounds not only for the solution of a single system (1) but also for a whole set of solutions − a problem arising by inexact input data or by enclosing the solution of a nonlinear system of equations.

Iterative processes in interval computation are known since more than two decades (cf. [4]). To a great part they imitate classical methods, but − due to interval arithmetic − they often show a different behaviour concerning

a) criteria of global convergence and
b) the speed of convergence.

In addition they induce the question of

c) permanence [1] and quality of enclosure of the iterates $[x]^k$.

In the sequel we will investigate these items for systems of linear equations of which the right hand sides vary in an interval vector $[b]$ and of which the

---

[1] i.e., $x=A^{-1}b\in[x]^0$ implies $x\in[x]^k, k=0, 1, \ldots$

coefficient matrices are M-matrices. If $[A]$ is splitted into $[A]=[M]-[N]$ we consider the iterative process

$$[x]^{k+1}=\mathrm{IGA}([M],[N][x]^k+[b]), \qquad k=0,1,\ldots, \tag{3}$$

which is built analogously to (2). Here $\mathrm{IGA}([B],[c])$ denotes the interval vector resulting from the interval Gaussian algorithm, applied to the interval matrix $[B]$ and the "right hand side" $[c]$. Apart from a slight modification this algorithm uses the same formulae as the classical Gaussian algorithm without changes of columns or rows. (See [3].) One can easily show that (3) contains many standard iterative methods of interval analysis such as interval Jacobi method, interval Gauß-Seidel method, interval relaxation methods. In case of convergence the iterates of (3) tend to a limit $[x]^*$ which encloses the solution set

$$S:=\{x\in\mathbb{R}^n\,|\,Ax=b \text{ for some } A\in[A], b\in[b]\}. \tag{4}$$

But the width of $[x]^*$ and the speed of convergence depend heavily on the splitting of $[A]$ (cf. Section 3 and 4). Therefore we will investigate method (3) for a special class of interval splittings which we called M-splittings in analogy to the classical case (cf. Section 2). For two of such splittings we will compare the corresponding sequences of iterates

a) with respect to the quality of enclosure,
b) with respect of the speed of convergence.

Section 3 and Section 4 are devoted to these two aspects. We will derive results similar to those for M-splittings of real matrices $A\in\mathbb{R}^{n\times n}$ described, e.g., in [20]. Section 5 illustrates our results by two simple examples. In order not to overload the paper with technical details we only report on the results while referencing their proofs.

2. Notations

We denote by $\mathrm{I}\mathbb{R}^n$, $\mathrm{I}\mathbb{R}^{n\times n}$ the set of real interval vectors with n components, and the set of real $n\times n$ interval matrices, respectively. We write interval quantities in square brackets, matrices in capital letters, vectors and scalars in small letters. Without further reference we use the notation $x=(x_i)$, $A=(a_{ij})$ for real vectors and real matrices. By $[\underline{A},\bar{A}]$, $([a]_{ij})$, $([\underline{a}_{ij},\bar{a}_{ij}])$ we mean the same interval matrix $[A]$ and we use a similar notation for interval vectors and intervals. Point intervals, i.e. degenerate intervals $[c,c]$, can be identified with the element, which they contain; therefore we write c instead of $[c,c]$.

For intervals $[a]$, $[b]$ we define the absolute value $|[a]|$, the width $d[a]$ and the distance $q([a],[b])$ by

$$|[a]|:=\max\{|\underline{a}|,|\bar{a}|\}, \quad d[a]:=\bar{a}-\underline{a} \quad \text{and} \quad q([a],[b]):=\max\{|\underline{a}-\underline{b}|,|\bar{a}-\bar{b}|\}.$$

For interval vectors and interval matrices these quantities are defined entrywise, e.g., $d[A]:=(d[a]_{ij})$ for $[A]\in\mathrm{I}\mathbb{R}^{n\times n}$; $d[A]$ is a nonnegative real $n\times n$ matrix where nonnegativity refers to the usual entrywise defined partial ordering $\leq$

with which we equip $\mathbb{R}^n$ and $\mathbb{R}^{n \times n}$. We extend this partial ordering to $I\mathbb{R}^{n \times n}$ by setting

$$[A] \leq [B] :\Leftrightarrow \underline{A} \leq \underline{B} \quad \text{and} \quad \bar{A} \leq \bar{B}.$$

We call a nonsingular matrix $A \in \mathbb{R}^{n \times n}$ an M-matrix, if its off-diagonal entries are not positive and if its inverse is nonnegative. Correspondingly we use the term M-matrix for an interval matrix $[A] \in I\mathbb{R}^{n \times n}$ if all its elements $A \in [A]$ are real M-matrices (cf. [5]).

Generalizing Schneider's definition of M-splittings in [18] mentioned above, we call $[A] = [M] - [N]$ an M-splitting of $[A] \in \mathbb{R}^{n \times n}$ if $[M]$ is an M-matrix and if $[N] \geq 0$.

Denote by $\rho(A)$ the spectral radius of $A \in \mathbb{R}^{n \times n}$ and let $P \in \mathbb{R}^{n \times n}$ be a fixed nonnegative matrix with $\rho(P) < 1$. If $f : I\mathbb{R}^n \rightarrow I\mathbb{R}^n$ is an interval function satisfying $q(f([x]), f([y])) \leq P q([x], [y])$ for all choices of $[x]$, $[y] \in I\mathbb{R}^n$, then we call f a P-contraction (see [3]). It is well known (cf. [3] or [12]) that a P-contraction has a unique fixed point $[x]^*$ and that the iterative process

$$[x]^{k+1} = f([x]^k), \qquad k = 0, 1, \ldots, \tag{5}$$

is globally convergent to $[x]^*$. Furthermore $\rho(P)$ is an upper bound for the R_1-factor of (5) which is defined as in [3] by

$$R_1 := \sup \left\{ \lim_{k \to \infty} \sup \| q([x]^k, [x]^*) \|^{1/k} \mid \{[x]^k\} \in \mathscr{C} \right\}$$

($\| \cdot \|$ any vector norm of $\mathbb{R}^n$; $\mathscr{C}$ set of all sequences constructed by (5) and therefore converging to $[x]^*$.)

This measure relates the speed of convergence of the sequence $\{\| q([x]^k, [x]^*) \|\}$ to that of the geometric sequence $\{R_1^k\}$. The value R_1 is independent of the norm $\| \cdot \|$ and of the starting vector $[x]^0$. In the sequel it always refers to method (3).

3. The Enclosure of the Solution Set S

Interval arithmetic shows quite a different algebraic behaviour than real arithmetic. The lack of inverses and distributivity often causes overestimations of sets and break downs of algorithms constructed by ideas of classical numerical analysis. Thus one of the first questions to answer is that of the feasibility of a method. E.g., it is well known that the interval Gaussian algorithm needed to define (3) may break down even if all elements of $[A]$ are nonsingular (cf. [17]). Conditions necessary and sufficient to assure its feasibility are still missing. Therefore we restrict ourselves to considering (3) under the following assumptions:

$$\begin{aligned} [A] &\in I\mathbb{R}^{n \times n} && \text{is an } M\text{-matrix}, \\ [A] &= [M] - [N] && \text{is an } M\text{-splitting of } [A], \\ [b] &\in I\mathbb{R}^n. \end{aligned} \tag{6}$$

These assumptions guarantee that the function

$$f: [x] \rightarrow f([x]) := \text{IGA}([M], [N][x] + [b])$$

is defined for each $[x] \in I\mathbb{R}^n$ (cf. [3]) and that it is a P-contraction. This forms the base of the following theorem allowing a first glance at the enclosure of the solution set S (defined in (4)) — the subject of this section.

Theorem 1: ([11], [12]) *Assuming* (6) *the following assertions hold*:

a) *Method* (3) *is feasible for any starting vector* $[x]^0 \in I\mathbb{R}^n$.
b) *All its sequences* $\{[x]^k\}$ *of iterates converge to the same interval vector* $[x]^*$ *which depends on* $[M]$ *and* $[N]$.
c) *If the solution set* S *is contained in* $[x]^0$ *then it is contained in each of the subsequent iterates.*
d) $S \subseteq [x]^*$.
e) $[x]^1 \subseteq [x]^0 \Rightarrow S \subseteq [x]^k, k = 0, 1, \dots$ □

Despite the global convergence of (3) and the enclosure of S by the limit $[x]^*$ there is no guarantee that S is contained in $[x]^k$ when starting (3) by an arbitrary $[x]^0 \in I\mathbb{R}^n$. This remark becomes essential when iterating on a computer since one has to stop after a finite number of steps, of course. It is the conditions in c) and e) which assure enclosure in each iterate. Fortunately one can often find starting vectors $[x]^0$ satisfying $S \subseteq [x]^0$ or $[x]^1 \subseteq [x]^0$. E.g., if u is a real vector such that

$$\underline{A}u \geq |[b]| \tag{7}$$

one gets $S \subseteq [-u, u]$. This follows at once by some properties of nonnegative matrices: (7) implies $u \geq \underline{A}^{-1}|[b]| \geq 0$, and $A \in [A]$, $b \in [b]$, $x = A^{-1}b \in S$ yield to

$$|x| = |A^{-1}b| \leq |A^{-1}||b| \leq \underline{A}^{-1}|[b]| \leq \underline{A}^{-1}|[b]| \leq u$$

which proves the assertion.

We point out, that vectors u satisfying (7) always exist: By a theorem of Fan [7] on M-matrices there are positive vectors v such that $\underline{A}v > 0$, hence u can be chosen as a suitable multiple of v. Vectors u can often be found by inspecting $\underline{A}$. For special M-matrices they are known in advance. E.g., let $\underline{A}$ be strictly diagonally dominant then $u := (1, 1, \dots, 1)^T$ satisfies $\underline{A}u > 0$.

In our second theorem we give conditions for $[x]^*$ to be the interval hull $[x]^H$ of S, i.e., the tightest enclosure of S by an interval vector. This is by no means trivial because the following example of Barth and Nuding shows that $[x]^* = \text{IGA}(A], [b])$ may overestimate $[x]^H$. Method (3) reduces to this direct method when choosing $[A] = [M]$ and $[N] = 0$.

Example: ([5])

Let

$$[A] = \begin{bmatrix} [\ 2, 4] & [-2, 0] \\ [-1, 0] & [\ 2, 4] \end{bmatrix}, \quad [b] = \begin{bmatrix} [\ 1, 2] \\ [-2, 2] \end{bmatrix}.$$

Then

$$\text{IGA}([A],[b])=\begin{bmatrix}[-1.5,4]\\[-2,\quad 3]\end{bmatrix}\gneqq[x]^H=\begin{bmatrix}[-1,\quad 4]\\[-1.5,3]\end{bmatrix}.$$

Theorem 2: ([5], [11], [15]) *Assuming* (6) *the limit* $[x]^*$ *of* (3) *equals the hull* $[x]^H$ *of S if one of the following conditions hold*:

  (i) $[M]$ *is a lower triangular matrix.*
 (ii) $[M]$ *is a point matrix.*
(iii) $0\leq[b]$ *or* $0\in[b]$ *or* $0\geq[b]$.   □

If $[A]$ is a point matrix then $\underline{A}=\bar{A}=:A$, and by $0=d[A]=d([M]-[N])=d[M]+d[N]$ (see [3]) $[M]$ and $[N]$ must be point matrices, too. Thus Theorem 2 guarantees in particular, that in the point case $A\equiv[A]$, $b\equiv[b]$ the iterates of (3) converge to the solution of $Ax=b$ even if one starts with an interval vector $[x]^0$.

Since for lower triangular matrices $[M]$ equation (3) is equivalent to

$$[x]_i^{k+1}=\left\{-\sum_{j=1}^{i-1}[m]_{ij}[x]_j^{k+1}+([N][x]^k+[b])_i\right\}/[m]_{ii},\quad i=1,\ldots,n \qquad (8)$$

(cf. [14] or [11]), the interval Jacobi method and Gauß-Seidel method are special cases of (3) which yield to the hull of S.

If $[M]$ contains nondegenerate intervals on both sides of its diagonal the limit $[x]^*$ of (3) may become coarser. This is stated in our next theorem where the limits with respect to two splittings are compared.

Theorem 3: ([12]) *Let* (6) *hold for two splittings* $[M]-[N]=[\hat{M}]-[\hat{N}]$ *of* $[A]$. *If*

$$[\hat{M}]\leq[M],\quad d[\hat{M}]\geq d[M] \qquad (9)$$

or, equivalently,

$$[\hat{N}]\leq[N],\quad d[\hat{N}]\leq d[N]$$

then $[x]^*\subseteq[\hat{x}]^*$ *for the limits* $[x]^*$, $[\hat{x}]^*$ *of* (3) *using* $[M]$, $[N]$ *and* $[\hat{M}]$, $[\hat{N}]$, *respectively.*   □

Condition (9) means that matrix $[\hat{M}]$ is shifted to the left compared with $[M]$ (overlapping allowed) while its width is increased.

Choosing $[\hat{M}]=[A]$ and, consequently, $[\hat{N}]\doteq0$, the conditions in (9) are fulfilled for each splitting $[M]-[N]$ satisfying (6). Since setting $[\hat{M}]=[A]$ reduces (3) to the interval Gaussian algorithm, this direct method involves the worst enclosure of S. This is stated in the following corollary.

Corollary 1: *Among all splittings of* $[A]$ *satisfying* (6) *the interval Gaussian algorithm leads to the worst enclosure of the solution set S.*   □

We will illustrate this result by examples in Section 5.

Theorem 2 and Theorem 3 suggest a triangular matrix $[M]$ as a good choice. Since all such matrices lead to the hull $[x]^H$ of S one can ask for those matrices effecting the fastest approach to $[x]^H$. This question will be investigated in our next section.

4. The Speed of Convergence

A classical result of Varga ([20], Thm. 3.15]) states that in the point case $A = M - N \equiv [A]$, $b \equiv [b]$ method (3) converges the faster the smaller N is. We will show that an analogous theorem also holds in the interval case. We begin by a theorem which already indicates that quality of enclosure and speed of convergence may be contrary. As "indicator" we use the spectral radius $\rho(\underline{M}^{-1}\bar{N})$ which is an upper bound of the R_1-factor of (3).

Theorem 4: ([11], [12]) *Assume* (6) *for the two splittings* $[M] - [N]$ *and* $[\hat{M}] - [\hat{N}]$. *Then the following assertions hold:*

a) $R_1 \leq \rho(\underline{M}^{-1}\bar{N})$. $\qquad\qquad\qquad\qquad\qquad\qquad\qquad\qquad\qquad$ (10)

b) If $[\hat{M}] \leq [M]$ *or* $[\hat{N}] \leq [N]$ *then* $\rho(\underline{\hat{M}}^{-1}\bar{\hat{N}}) \leq \rho(\underline{M}^{-1}\bar{N})$. $\quad\square$ $\qquad$ (11)

Despite the fact that in the non-interval case $\rho(\underline{M}^{-1}\bar{N})$ equals the R_1-factor (cf. [16]), this needs not be true in the interval case.

Thus for $[A] = \begin{bmatrix} 1 & -[0,\frac{1}{2}] \\ -[0,\frac{1}{2}] & 1 \end{bmatrix}$ and $[b] = \begin{bmatrix} 1 \\ -1 \end{bmatrix}$ one can show (cf. [11]) that the R_1-factor of the Jacobi method is zero while $\rho(\underline{M}^{-1}\bar{N}) = \frac{1}{2}$. Therefore part b) of Theorem 4 must be considered with some caution although all our numerical experiments have confirmed its contents.

If $[M]$ is lower triangular, a necessary and sufficient condition is known for equality to hold in (10) (cf. [13]). As a conclusion of this condition we mention the following two cases:

1) If $d[A] = 0$, i.e. if $[A]$ is a point matrix, then $R_1 = \rho(\underline{M}^{-1}\bar{N})$ holds.
2) If $d[A] > 0$, i.e. if each entry of $[A]$ is a nondegenerate interval, then equality holds in (10) if and only if

$$\underline{A}^{-1}\underline{b} \leq 0 \quad \text{or} \quad \underline{A}^{-1}\bar{b} \geq 0. \qquad\qquad\qquad (12)$$

In all cases of triangular matrices $[M]$ one can replace the spectral radii in (11) by the corresponding R_1-factors.

Theorem 5: ([11], [13]) *Assume* (6) *for the two splittings* $[M] - [N]$ *and* $[\hat{M}] - [\hat{N}]$ *and let* R_1, $\hat{R}_1$ *be the corresponding R_1-factors of* (3). *If A is irreducible and if* $[M]$, $[\hat{M}]$ *are lower triangular matrices satisfying* $[\hat{M}] \leq [M]$ (*which is equivalent to* $[\hat{N}] \leq [N]$) *then* $\hat{R}_1 \leq R_1$. $\quad\square$

Since equality holds in (10) for the point case Theorem 5 generalizes Varga's theorem mentioned at the beginning of this section. If $[\hat{M}]$ is the lower triangular matrix of the Gauß-Seidel splitting then it is easily seen that $[\hat{M}] \leq [M]$ holds for each admissible matrix $[M]$ of Theorem 5. This reveals at once the special role of the interval Gauß-Seidel method:

Corollary 2: ([11], [13]) *Let the assumptions* (6) *hold. Then the R_1-factor is small-est for the Gauß-Seidel method when compared with R_1 of splittings $[A]=[M]-[N]$ in which $[M]$ if lower triangular.* □

Combining Theorem 2 with Corollary 2 the Gauß-Seidel iteration turns out to be a very favourable iterative method. This will also be confirmed in our examples in Section 5.

We already mentioned that $R_1=\rho(M^{-1}N)$ in the point case $A=M-N\equiv[A]$, $b\equiv[b]$. Thus the famous Theorem of Stein and Rosenberg (cf. [20], e.g.) can be reformulated in terms of R_1-factors, if A is an M-matrix. Doing it there is an analogue of this theorem in interval analysis, containing it as a special case.

Theorem 6: (Theorem of Stein and Rosenberg for interval matrices; [11]) *Let $[b]\in I\mathbb{R}^n$ and let $[A]\in I\mathbb{R}^{n\times n}$ be an M-matrix with irreducible lower bound $\underline{A}$. Then the R_1-factor R_1^J and R_1^S of the Jacobi method and of the Gauß-Seidel method, respectively, fulfill the following alternative:*

$$\text{either} \quad R_1^S=R_1^J=0 \quad \text{or} \quad 0<R_1^S<R_1^J<1. \quad \square$$

This theorem again emphasizes the exposed role of Gauß-Seidel iteration.

5. Examples

In this section we illustrate the results of the preceding sections by two examples. We begin by a very simple one.

Example 1:

Let

$$[A]:=\begin{pmatrix} [\ 2,3] & [-1,0] & [-1,0] \\ [-1,0] & [\ 4,6] & [-1,0] \\ [-1,0] & [-2,0] & [\ 6,8] \end{pmatrix}$$

and

$$[b]:=([1,2],[1,2],[-2,-1])^T.$$

Since $Ax>0$ for $x:=(2,1,1)^T$ and for each $A\in[A]$, the matrices A are M-matrices by a criterion of Fan [7]. Using, e.g., Jacobi method and taking Theorem 2 into account one easily sees that

$$[x]^H:=([\tfrac{2}{9},\tfrac{55}{31}],[\tfrac{1}{9},\tfrac{33}{31}],[-\tfrac{1}{3},\tfrac{15}{31}])^T$$

is the interval hull of the solution set S associated with $[A]$ and $[b]$. We split $[A]$ into $[M]-[N]$ by setting either $[m]_{ij}:=[a]_{ij}$ or $[m]_{ij}:=0$ depending on i and j. Then $[N]$ is uniquely defined by $[A]=[M]-[N]$. We iterate by using several splittings, indicating the choice $[m]_{ij}=[a]_{ij}$ symbolically by "x" in the following table. As can be easily seen, the inequalities $[M]_1\leq[M]_2$, $d[M]_1\geq d[M]_2$ hold for two successive splittings in Table 1 until $[M]_2$ reaches diagonal form; from that point on $[M]_1\geq[M]_2$, $d[M]_1\leq d[M]_2$. The second column lists the numbers k_s necessary to fulfill the stopping criterion

$$\left.\begin{array}{l}|\underline{x}_i^k - \underline{x}_i^{k-1}| \le 10^{-6} \cdot |\underline{x}_i^{k-1}| \\ |\bar{x}_i^k - \bar{x}_i^{k-1}| \le 10^{-6} \cdot |\bar{x}_i^{k-1}|\end{array}\right\} i = 1, 2, 3,$$

when starting with $[x]^0 = (0, 0, 0)^T$. These numbers k_s reflect the speed of convergence in accordance with the upper bound $\rho(\underline{M}^{-1}\bar{N})$ of the R_1-factor, for which we list approximations in column 3 computed by using the characteristic polynomial and Newton's method. By (12) this bound equals R_1 at least for $[M]$ being lower triangular. In the columns 4−6 we list the lower bounds $\underline{x}_i^{k_s}$, $i = 1$,

Table 1

$[M]$	k_s	$\rho(\underline{M}^{-1}\bar{N})$	$\underline{x}_1^{k_s}$	$\underline{x}_2^{k_s}$	$\underline{x}_3^{k_s}$
× × × × × × × × ×	1	0	0.1827	0.0537	−0.4517
× × ○ × × × × × ×	9	0.1622	0.2072	0.1036	−0.3784
× × ○ × × ○ × × ×	12	0.2619	0.2222	0.1111	−0.3334
× ○ ○ × × ○ × × ×	16	0.3871	0.2222	0.1111	−0.3334
× ○ ○ × × ○ ○ × ×	21	0.5000	0.2222	0.1111	−0.3334
× ○ ○ ○ × ○ ○ × ×	26	0.5781	0.2222	0.1111	−0.3334
× × ○ ○ × ○ ○ ○ ×	30	0.6257	0.2222	0.1111	−0.3334
× × ○ ○ × ○ ○ ○ ×	26	0.5761	0.2222	0.1111	−0.3334
× × ○ ○ × × ○ ○ ×	23	0.5326	0.2222	0.1111	−0.3334
× × ○ × × × ○ × ×	15	0.3663	0.2105	0.1052	−0.3685
× × × × × × ○ × ×	9	0.1389	0.2037	0.0694	−0.3889

2, 3; the upper ones coincide in all cases with $\bar{x}_i^H$ (within rounding):

$$\bar{x}_1^{k_s} = 1.775, \qquad \bar{x}_2^{k_s} = 1.065, \qquad \bar{x}_3^{k_s} = 0.4839.$$

We made our calculations on a kws SAM 68 K computer using a floating point system of mantissa length 12. The programs were written in the language PAS-CAL-SC (cf. [10]), an extension of standard PASCAL which is based on an accurate scalar product and on a machine interval arithmetic as described, e.g., in [3] and [9]. Therefore we may suppose that the values of $[x]^{k_s}$ are those of the limit $[x]^*$.

The results in Table 1 confirm all our theorems stated in the preceding two sections. In particular they show that an (upper or lower) triangular matrix $[M]$ leads to the hull of the solution set S while non-triangular matrices $[M]$ can overestimate $[x]^H$. They also demonstrate the special role of Gauß-Seidel iteration among splittings with $[M]$ being triangular. Using the representation (8) for iterative processes based on such splittings, one easily sees that the amount of work per step of iteration remains the same. This shows again that k_s, R_1-factor and $\rho(M^{-1}\bar{N})$ are realistic measures for the speed of convergence when $[M]$ is triangular. Comparing the interval Gaussian algorithm to Gauß-Seidel method with respect to the total amount of multiplications/divisions (which equals approximately that of additions/subtractions), Gauß-Seidel method is inferior in this example, the relation being 17 to $16 \cdot 9$. This relation gets reversed if the dimension and the number of zero entries of $[A]$ increase. We demonstrate this phenomenon in our second example:

Example 2:

Let $[A]$ be a 100×100 five-diagonal matrix defined by

$$[a]_{ij} := \begin{cases} 8, & \text{if } j=i, \\ [-2,0], & \text{if } j=i+1 \quad \text{or} \quad j=i+20, \\ [-1,0], & \text{if } j=i-1 \quad \text{or} \quad j=i-20, \\ 0, & \text{otherwise}, \end{cases}$$

and set $[b]_i := [-1, 1]$, $i = 1, \ldots, 100$.

Again $[A]$ is an M-matrix since $Ax > 0$ for $x = (1, 1, \ldots, 1)^T \in \mathbb{R}^{100}$ and for any $A \in [A]$ (cf. [7]). We construct $[M]$ by replacing some of the five non-zero diagonals

$$\{[a]_{ij} \,|\, j = i+d\}, \qquad d \in \{-20, -1, 0, 1, 20\} \tag{13}$$

of $[A]$ completely by zero entries. In Table 2 we indicate the non-zero diagonals of $[M]$ by listing the associated integers d of (13), e.g., -20, -1, 0 for Gauß-Seidel method. By Theorem 2 each of our splittings yields to the hull of the solution set S. Therefore we can concentrate our attention to the speed of convergence and its comparison with the total amount of interval multiplications/divisions. We use the same stopping criterion as in Example 1 denoting again the number of iterates by k_s. This time we compute $\rho(M_k^{-1}\bar{N})$ by the von-Mises

Table 2

$[M]$	k_s	$\rho(\underline{M}^{-1}\bar{N})$	Approximate total amount of interval multiplications/ divisions $(n=100;\ \beta=20)$
$-20, -1, 0, 1, 20$	1	0	$(n-2\beta)(\beta+1)^2 = 26460$
$-20, -1, 0, 1$	13	0.23413	$k_s \cdot [3(n-2\beta)(\beta+1)] = 49140$
$-20, -1, 0$	25	0.53266	$k_s \cdot (5n) = 12500$
$-1, 0$	31	0.63055	$k_s \cdot (5n) = 15500$
0	35	0.67762	$k_s \cdot (5n) = 17500$
$0, 1$	26	0.57337	$k_s \cdot (5n) = 13000$
$-1, 0, 1$	21	0.48696	$k_s \cdot (7n) = 14700$
$-1, 0, 1, 20$	12	0.24049	$k_s \cdot [2(n-2\beta)(\beta+2)] = 31680$

method and we estimate the numbers of operations roughly taking (8) and the band structure of $[M]$ into account.

Again the results confirm the theorems previously stated. In addition, Table 2 displays Gauß-Seidel method being the best one due to the sparsity of $[A]$ and the fill-in of the band when using the interval Gaussian algorithm.

6. Conclusions

The examples in Section 5 as well as Theorem 2 in Section 3 show that iterative processes (3) result in the tightest enclosure of the solution set S if $[M]$ is chosen as triangular matrix. Among those matrices that of Gauß-Seidel method gives the smallest R_1-factor. Therefore it seems natural to iterate by this method. However, when taking the total amount of arithmetic operations into account (and also in view of Theorem 4) splittings with non-triangular matrices $[M]$ may be superior involving the dilemma of being fast but not necessarily tightly enclosing. In the case of large sparse systems of linear equations Gauß-Seidel method normally turns out to be the best method, non-triangular matrices $[M]$ often causing fill-in and thus requiring a larger amount of storage.

References

[1] Albrecht, J.: Monotone Iterationsfolgen und ihre Verwendung zur Lösung linearer Gleichungssysteme. Numer. Math. 3, 345−358 (1961).
[2] Alefeld, G.: Über die aus monoton zerlegbaren Operatoren gebildeten Iterationsverfahren. Computing 6, 161−172 (1970).
[3] Alefeld, G., Herzberger, J.: Introduction to interval computations. Academic Press, New York, 1983.
[4] Apostolatos, N., Kulisch, U.: Grundzüge einer Intervallrechnung für Matrizen und einige Anwendungen. Elektron. Rech. 10, 73−83 (1968).
[5] Barth, W., Nuding, E.: Optimale Lösung von Intervallgleichungssystemen. Computing 12, 117−125 (1974).
[6] Collatz, L.: Funktionalanalysis und numerische Mathematik. Springer Verlag, Berlin, 1964.
[7] Fan, K.: Topological proofs for certain theorems on matrices with non-negative elements. Monatsh. Math. 62, 219−237 (1958).

G. Mayer

[8] Hackbusch, W.: Theorie und Numerik elliptischer Differentialgleichungen. Teubner Verlag, Stuttgart, 1986.
[9] Kulisch, U., Miranker, W.L.: Computer arithmetic in theory and practice. Academic Press, New York, 1981.
[10] Kulisch, U., Miranker, W.L.: A new approach to scientific computation. Academic Press, New York, 1983.
[11] Mayer, G.: Reguläre Zerlegungen und der Satz von Stein und Rosenberg für Intervallmatrizen. Habilitationsschrift. Karlsruhe, 1986.
[12] Mayer, G.: Comparison theorems for iterative methods based on strong splittings. SIAM J. Numer. Anal. *24*, 215 – 227 (1987).
[13] Mayer, G.: On the speed of convergence of some iterative processes. In: Colloquia Mathematica Societatis János Bolyai, Vol. 50. Numerical Methods. Miskolc, 1986. North Holland, Amsterdam, 200 – 221 (1988).
[14] Neumaier, A.: New techniques for the analysis of linear interval equations. Linear Algebra Appl. *58*, 273 – 325 (1984).
[15] Neumaier, A.: Further results on linear interval equations. Linear Algebra Appl. *87*, 155 – 179 (1987).
[16] Ortega, J.M., Rheinboldt, W.C.: Iterative solution of nonlinear equations in several variables. Academic Press, New York, 1970.
[17] Reichmann, K.: Abbruch beim Intervall-Gauß-Algorithmus. Computing *22*, 355 – 361 (1979).
[18] Schneider, H.: Theorems on *M*-splittings of a singular *M*-matrix which depend on graph structure. Linear Algebra Appl. *58*, 407 – 424 (1984).
[19] Schröder, J.: Fehlerabschätzungen bei linearen Gleichungssystemen mit dem Brouwerschen Fixpunktsatz. Arch. Rational Mech. Anal. *3*, 28 – 44 (1959).
[20] Varga, R.S.: Matrix iterative analysis. Prentice-Hall, Englewood Cliffs, N.J., 1962.

Priv.-Doz. Dr. Günter Mayer
Institut für Angewandte Mathematik
Universität Karlsruhe
Kaiserstrasse 12
D-7500 Karlsruhe 1
Federal Republic of Germany

Computing, Suppl. 6, 59–68 (1988)

Errorbounds for Quadratic Systems of Nonlinear Equations Using the Precise Scalar Product

G. Alefeld, Karlsruhe

Abstract — Zusammenfassung

Errorbounds for Quadratic Systems of Nonlinear Equations Using the Precise Scalar Product. For nonlinear systems of quadratic equations we show how the precise scalar product can be used in order to compute and to improve inclusions for a solution. Our main interest is the special case which comes from the generalized eigenvalue problem.

Fehlerschranken für quadratische Systeme nichtlinearer Gleichungen unter Verwendung des genauen Skalarprodukts. Für Gleichungssysteme des angegebenen Typs berechnen wir Einschließungen für eine Lösung und verbessern diese unter Verwendung des genauen Skalarprodukts. Das Hauptinteresse gilt dem Spezialfall, welcher dem verallgemeinerten Eigenwertproblem entspricht.

1. Introduction

We consider the *quadratic equation*

$$y = u + S y + T y^2 \tag{1}$$

where $u = (u_i)$ is a real vector from $\mathbb{R}^n$, $S = (s_{ij})$ is a real (n, n) matrix and $T = (t_{ijk})$ is a real bilinear operator from $\mathbb{R}^n \times \mathbb{R}^n$ to $\mathbb{R}^n$ defined by $Tx\, y = \left(\sum_{j=1}^{n} \sum_{k=1}^{n} t_{ijk} x_k\, y_j \right)$ for $x = (x_i)$, $y = (y_i) \in \mathbb{R}^n$. Ty^2 is defined to be $Ty\, y$. The unknown vector is $y = (y_i)$. In general a solution of (1) can only be computed approximately by an iterative method. Furthermore if such a method is performed on a computer one has to take into account rounding errors. We show how this can be done for the quadratic (1) using the precise scalar product.

Our main interest is a special case of (1), namely the generalized eigenvalue problem. Note that quadratic equations have already been considered in [1]. The special case of an eigenvalue problem has been discussed in [2] and [3].

2. The Generalized Eigenvalue Problem as a Quadratic Equation of the Form (1)

We consider the generalized matrix eigenvalue problem

$$A x = \lambda B x \tag{2}$$

where A and B are real (n, n) matrices. We assume in this paper that B is nonsingular. Under practical aspects this is not very restrictive since in real life problems B is usually even symmetric and positive definite.

Assume now that after using some well known algorithm for computing eigen-pairs (see [9], for example) we have given a real approximation λ for a simple real eigenvalue and an approximation x for the corresponding eigenvector. For the exact eigenpair $(\lambda + \tilde{\mu}, x + \tilde{y})$ the equation

$$A(x + \tilde{y}) = (\lambda + \tilde{\mu}) B(x + \tilde{y}) \tag{3}$$

holds. Let

$$\|x\|_\infty = |x_s| > 0 \tag{4}$$

where s is some index for which the infinity norm is taken on. Since $x + \tilde{y}$ is not unique we normalize $x + \tilde{y}$ by setting

$$\tilde{y}_s = 0 \tag{5}$$

where $\tilde{y} = (\tilde{y}_i)$. Equation (3) can be rewritten as

$$(A - \lambda B)\,\tilde{y} - \tilde{\mu} B x = (\lambda B - A)\,x + \tilde{\mu} B \tilde{y}. \tag{6}$$

Defining the components y_i of the vector $y = (y_i) \in \mathbb{R}^n$ by

$$y_i = \begin{cases} \tilde{y}_i, & i \neq s \\ \tilde{\mu}, & i = s \end{cases}$$

the last equation can be written as

$$C y = r + B(y_s\,\tilde{y}) \tag{7}$$

where

$$r = \lambda B x - A x \tag{8}$$

and where C is identical to $A - \lambda B$ with the exception of the s-th column which is replaced by $-Bx$. See [5].

If B is nonsingular — this was our general assumption — and for sufficiently good approximations λ and x it can be shown that the matrix C is nonsingular (see [8], for example). Assume now that this is the case and let L be some approximation to the inverse of C. Then (7) can be rewritten as

$$y = L r + (I - LC)\,y + L(B(y_s\,\tilde{y})). \tag{9}$$

This equation has the form (1) where $u = L r$ and $S = I - LC$. The bilinear operator T is defined by $T y^2 = L(B y_s\,\tilde{y}))$. We omit to express the elements t_{ijk} of T explicitly by the elements of the matrices L and B because this is not important in the sequel. Note, however, that the last term in (9) could also be written as

$$L(B(y_s\,\tilde{y})) = (LB)(y_s\,\tilde{y}) = y_s(LB)\,\tilde{y}$$

since the associative law holds and since y_s is a scalar. The reason why we use the first one of these equivalent expressions becomes clear in Chapters 3 and 4.

3. Computing an Enclosing Interval Vector

We now try to compute an interval vector $[y] = ([y]_i)$ for which

$$u + Sy + Ty^2 \in [y] \quad \text{for all } y \in [y]. \tag{10}$$

Then by the Brouwer fixed-point theorem the equation (1) has at least one solution in $[y]$. We try to find $[y]$ in the form

$$[y] = [-\beta, \beta] e \tag{11}$$

where $\beta > 0$ and $e = (1, 1, \ldots, 1)^T \in \mathbb{R}^n$ (This approach is motivated by the fact that $y = 0$ is nearly a solution of (1) if u is "small"). By inclusion monotonicity (see [4], Chapter 1, Theorem 5) we have for $y \in [y]$

$$u + Sy + Ty^2 \in u + S[y] + T[y]^2.$$

Hence

$$[w] := u + S[y] + T[y]^2 \subseteq [y] \tag{12}$$

is sufficient for (10).

Following the laws of interval arithmetic we have

$$S[y] = \left(\sum_{j=1}^{n} s_{ij} [-\beta, \beta] \right) = [-\beta, \beta] \left(\sum_{j=1}^{n} |s_{ij}| \right)$$

and

$$T[y]^2 = \left(\sum_{j=1}^{n} \left(\sum_{k=1}^{n} t_{ijk} y_k \right) y_j \right)$$

$$= \left(\sum_{j=1}^{n} \left(\sum_{k=1}^{n} t_{ijk} [-\beta, \beta] \right) [-\beta, \beta] \right)$$

$$= [-\beta^2, \beta^2] \left(\sum_{j=1}^{n} \left(\sum_{k=1}^{n} |t_{ijk}| \right) \right)$$

and therefore

$$u + Sy + Ty^2 = u + [-\beta, \beta]|S|e + [-\beta^2, \beta^2]|T|e^2$$

where $|S| = (|S_{ij}|)$ and $|T| = (|t_{ijk}|)$.

(12) holds iff

$$|m[w] - m[y]| + \tfrac{1}{2} d[w] \leq \tfrac{1}{2} d[y] \tag{13}$$

where m denotes the center, d the diameter and $|\cdot|$ the absolute value of an interval vector (see [4], Chapter 10).

We have

$$m[w] = u, \quad m[y] = 0,$$
$$d[w] = 2\beta|S|e + 2\beta^2|T|e^2,$$
$$d[y] = 2\beta e.$$

Hence (13) holds iff

$$|u| + \beta|S|e + \beta^2|T|e \leq \beta e. \tag{14}$$

Defining

$$\rho = \| u \|_\infty, \tag{15}$$

$$\kappa = \| S \|_\infty, \tag{16}$$

$$\ell = \max_i \left\{ \sum_{j=1}^n \sum_{k=1}^n |t_{ijk}| \right\} \tag{17}$$

the last vector inequality holds if

$$\rho + \beta \kappa + \beta^2 \ell \le \beta.$$

(For the bilinear operator $T = (t_{ijk})$ a norm can be defined by

$$\| T \|_\infty = \sup_{\| x \|_\infty = \| y \|_\infty = 1} \| T x y \|_\infty.$$

It is easy to prove that $\| T \|_\infty \le \ell$ where ℓ is defined by (17). However, in general equality does not hold. See [7], for example).

Hence we have the following result.

Theorem 1. *Let* ρ, κ, ℓ *be defined by* (15)−(17). *Assume that* $\kappa < 1$, $(1-\kappa)^2 - 4\rho\ell \ge 0$ *and let*

$$\beta_{1/2} = \frac{1 - \kappa \mp ((1-\kappa)^2 - 4\rho\ell)^{1/2}}{2\ell} \tag{18}$$

be the solutions of the quadratic equation

$$\beta^2 \ell + (\kappa - 1)\beta + \rho = 0. \tag{19}$$

If $\beta \in [\beta_1, \beta_2]$ *then the equation* (1) *has at least one solution* y^* *in the interval vector* (11). □

Please note that Theorem 1 gives only sufficient conditions for (10).

In the special case of the equation (9) the preceding theorem holds for

$$\rho = \| Lr \|_\infty \tag{20}$$

$$\kappa = \| I - LC \|_\infty \tag{21}$$

$$\ell = \| \, |L| |B| \, \|_\infty. \tag{22}$$

If we choose $L := C^{-1}$ in (9) then the equation (9) reads

$$y = C^{-1} r + C^{-1}(B(y_s, \tilde{y})) \tag{23}$$

and the condition (10) can be written as

$$C^{-1}(r + B([y]_s, [\tilde{y}])) \subseteq [y] \tag{24}$$

(Note that for a real matrix M and interval vectors $[x]$ and $[y]$ it holds that $M([x] + [y]) = M[x] + M[y]$).

If we would have written the equation (9) in the form

$$y = C^{-1} r + (C^{-1} B)(y_s, \tilde{y})$$

then we could not rewrite the condition (10) in the form (24) since in the set of interval vectors and interval matrices the associative law does not hold in general.

4. Improving an Enclosing Interval Vector Iteratively

After applying the preceding theorem one can try to improve the computed inclusion using the following iteration method:

$$\left.\begin{array}{l} [y]^0 = [-\beta, \beta]e \\ [y]^{k+1} = g([y]^k), \quad k = 0, 1, 2, \dots \end{array}\right\} \tag{25}$$

where

$$g[y] = u + S[y] + T[y]^2. \tag{26}$$

This iteration method computes a sequence $\{[y]^k\}_{k=0}^{\infty}$ of interval vectors for which the following result holds.

Theorem 2. *Let* $\kappa < 1$, $(\kappa - 1)^2 - 4\rho\ell > 0$ *and let* β_1, β_2 *be defined by* (18). *Then if*

$$\beta_1 \leq \beta < \frac{\beta_1 + \beta_2}{2} \tag{27}$$

it holds that

a) $y^* \in [y]^k$

and

b) $\lim_{k \to \infty} [y]^k = y^*$.

Furthermore y^* *is unique in* $[y]^0$.   □

We omit the details of a proof.

Please note again that the result also holds for

$$g([y]) = Lr + (I - LC)[y] + L(B([y]_s, [\tilde{y}])) \tag{28}$$

if ρ, κ and ℓ are defined by (20)–(22).

In the special case $L := C^{-1}$ we can rewrite (28) as

$$\begin{aligned} g([y]) &= C^{-1}r + C^{-1}(B[y]_s [\tilde{y}]) \\ &= C^{-1}(r + B([y]_s [\tilde{y}])). \end{aligned} \tag{29}$$

At first glance it does not make much sense to set $L := C^{-1}$ since normally C^{-1} can not be represented exactly on a computer. However, there are some good reasons why this makes sense. In the first place the term $(I - LC)[y]$ becomes zero, which means that it has not to be computed in every iteration step. Furthermore the fact that C^{-1} can not be represented exactly has not to bother us. It is well known that there exist algorithms using the precise scalar product which deliver an optimal inclusion of C^{-1}. Hence we only have to care for the accurate computation of $r + B([y]_s [\tilde{y}])$ in each iteration step. How this can be done is described in Chapter 6 for the general quadratic (1).

5. A Heuristic Procedure for Computing an Enclosing Interval Vector

Assume that the assumptions of Theorem 1 hold such that, using this theorem, we can compute an including interval vector for a solution of (1). From a practical point of view one is interested in a very good inclusion. Therefore the choice $\beta := \beta_1$ suggests itself. However, as the proof of Theorem 1 shows this choice is only sufficient for $g([y]^0) \subseteq [y]^0$. One could try to find $[y]^0$ with a smaller diameter such that $g([y]^0) \subseteq [y]^0$ by the following iteration method

$$
\begin{aligned}
&\text{Set } [z]^0 := 0; \\
&[z]^{k+1} := \text{conv}\{u + S[z]^k + T([z]^k)^2, [z]^k\} \\
&\qquad \text{until } [z]^{k+1} \subseteq [z]^k; \\
&[y]^0 := [z]^k;
\end{aligned}
\tag{30}
$$

where $\text{conv}\{\cdot, \cdot\}$ denotes the convex hull of two interval vectors.

By inclusion monotonicity of interval arithmetic we have $[z]^k \subseteq [-\beta_1, \beta_1]e$ for all k provided Theorem 1 applies.

Practical experience shows that applying Theorem 1 and the iteration method (30), respectively, nearly needs the same computing time. Using (30), however, has the advantage that it may still work if Theorem 1 is not applicable.

Of course these remarks hold also in the special case of the equation (9) which comes from the generalized eigenvalue problem (2).

6. Using the Precise Scalar Product

In order to get inclusions with small diameters on a computer one should try to keep the rounding errors as small as possible when performing the iteration method (25).

Our quadratic equation is an example for which the so-called precise scalar product can be applied (see [6]) to achieve this.

We show that the components of the right-hand side of the equation

$$
y = u + Sy + Ty^2
$$

can be computed by a single scalar product.

The only problem which has to be explained is how to reduce Ty^2 to one scalar product. This can be done in the following manner:

We have by definition

$$
Ty^2 = \left(\sum_{j=1}^{n} \left(\sum_{k=1}^{n} t_{ijk} y_k \right) y_j \right).
$$

Define now

$$r_{ijk} = f\ell(t_{ijk}\, y_k)$$

where $f\ell(\cdot)$ denotes the floating point multiplication.

Furthermore let

$$p_{ijk} = \begin{pmatrix} t_{ijk} \\ t_{ijk} \end{pmatrix}^T \cdot \begin{pmatrix} y_k \\ -1 \end{pmatrix}$$

be the precise scalar product. Then it holds exactly

$$t_{ijk}\, y_k = r_{ijk} + p_{ijk}.$$

Therefore we can write

$$Ty^2 = \left(\sum_{j=1}^{n} \left(\sum_{k=1}^{n} (r_{ijk} + p_{ijk}) \right) y_j \right)$$

$$= \left(\sum_{j=1}^{n} \left(\sum_{k=1}^{n} r_{ijk}\, y_j + \sum_{k=1}^{n} p_{ijk}\, y_j \right) \right)$$

and our problem is solved. Note that when performing the iteration method (25) we have interval vectors on the right-hand side. In this case a similar idea can be applied.

If we specialize our quadratic equation (1) to the eigenvalue problem (2) then u, S and T are in general not exactly representable on the machine and the situation becomes much more complicated. We do not discuss the general case but only the choice $L := C^{-1}$. As we have seen at the end of Chapter 4 we have in this case

$$g[y] = C^{-1}(r + B([y]_s\, [\tilde{y}])).$$

Using the same ideas as before the components of the expression $r + B([y]_s\, [\tilde{y}])$ can be computed each by one scalar product. We do not repeat the details.

We close with a final comment. Since it is clear that in general we can not compute C^{-1} exactly it seems to be more favourable to perform the iteration method

$$[y]^{k+1} = \mathrm{IGA}(C, r + B([y]_s^k\, [\tilde{y}]^k)),$$
$$k = 0, 1, 2, \ldots$$

where $\mathrm{IGA}(\cdot, \cdot)$ denotes the result of the Gaussian algorithm. Note, however, that for a real (n, n) matrix M and an interval vector $[b]$ one can prove that

$$M^{-1}[b] \subseteq \mathrm{IGA}(M, [b]).$$

See [4], Chapter 15. If one assumes that this also holds if rounding errors are taken into account — this has not been proved — then it is clear that the iteration method (25) with $g[y]$ defined by (29) has to be preferred in order to get narrow inclusions.

7. Numerical Examples

Consider the symmetric matrices

$$A = \begin{bmatrix} 12 & 1 & -1 & 2 & 1 \\ 1 & 14 & 1 & -1 & 1 \\ -1 & 1 & 16 & -1 & 1 \\ 2 & -1 & -1 & 12 & -1 \\ 1 & 1 & 1 & -1 & 11 \end{bmatrix}$$

and

$$B = \begin{bmatrix} 10 & 2 & 3 & 1 & 1 \\ 2 & 12 & 1 & 2 & 1 \\ 3 & 1 & 11 & 1 & -1 \\ 1 & 2 & 1 & 9 & 1 \\ 1 & 1 & -1 & 1 & 15 \end{bmatrix}$$

from [9], p. 313.

As λ we choose the first six digits of an approximation to an eigenvalue given in [9]:

$$\lambda = 0.231060 \times 10^{+1}.$$

Analogously we choose the corresponding eigenvector approximation

$$x = \begin{pmatrix} -0.204587 \\ 0.931721 \times 10^{-1} \\ 0.240022507111 \\ -0.166395 \\ 0.630418 \times 10^{-1} \end{pmatrix}.$$

(The third component which has the largest absolute value is exactly the approximation given in [9]. This component is not changed by our algorithm).

Setting $C = L^{-1}$ we get for β_1 from Theorem 1

$$\beta_1 = 0.4321587418763 \times 10^{-5}.$$

After two iteration steps of method (25) with g defined by (29) we get the following inclusions for the eigenvalue $\lambda + \tilde{\mu}$ and for the components of the corresponding eigenvector $x + \tilde{y}$:

$$\lambda + \mu \in [0.231060432134^8_9 \times 10^1]$$

$$x + \tilde{y} \in \begin{pmatrix} [-0.204586718184^5_4] \\ [0.931720977435^4_5 \times 10^{-1}] \\ [0.240022507111\,0] \\ [-0.166395354479^8_7] \\ [0.630417653106^7_8 \times 10^{-1}] \end{pmatrix}$$

The computation of β_1 needs 6063 ms. The overall computing time is 10022 ms. Using (30) we have $[z]^{k+1} \subseteq [z]^k$ after 4 steps. This needs 7630 ms. After three iteration steps of (25) with (29) we get the same final inclusions. The overall computing time is in this case 10012 ms.

Practical experience shows that for good approximations λ and x both approaches need nearly the same computing time. If, however, Theorem 1 is not applicable then the second approach still works if the approximations are not too bad.

References

[1] Alefeld, G.: Componentwise Inclusion and Exclusion Sets for Solutions of Quadratic Equations in Finite Dimensional Spaces, Numer. Math. 48, 391–416 (1986)
[2] Alefeld, G.: Berechenbare Fehlerschranken für ein Eigenpaar unter Einschluß von Rundungsfehlern bei Verwendung des genauen Skalarprodukts. Z. angew. Math. Mech. 67, 3, 145–152 (1987).
[3] Alefeld, G.: Berechenbare Fehlerschranken für ein Eigenpaar beim verallgemeinerten Eigenwertproblem. Z. angew. Math. Mech. 68, 3, 181 ff. (1988).
[4] Alefeld, G., Herzberger, J.: Introduction to Interval Computations. Academic Press, New York, 1983 (ArNu).
[5] Dongarra, J.J., Moler, C.B., Wilkinson, J.: Improving the accuracy of computed eigenvalues and eigenvectors. SIAM J. Numer. Anal. 20, 23–45 (1983).
[6] Kulisch, U., Miranker, W. (Eds.): A New Approach to Scientific Computation. Academic Press, 1983.
[7] Platzöder, L.: Einige Beiträge über die Existenz von Lösungen nichtlinearer Gleichungssysteme und Verfahren zu ihrer Berechnung. Dissertation. Technische Universität Berlin (1981).
[8] Symm, H.J., Wilkinson, J.H.: Realistic error bounds for a simple eigenvalue and its associated eigenvector. Numer. Math. 35, 113–126 (1980).
[9] Wilkinson, J.H., Reinsch, C.: Handbook for Automatic Computation. Volume 2: Linear algebra. Springer Verlag (1971).

Prof. Dr. G. Alefeld
Institut für Angewandte Mathematik
Universität Karlsruhe
Kaiserstrasse 12
D-7500 Karlsruhe 1
Federal Republic of Germany

Computing, Suppl. 6, 69–78 (1988)

Computing

© by Springer-Verlag 1988

Inclusion of Eigenvalues of General Eigenvalue Problems for Matrices

H. Behnke, Clausthal

Abstract — Zusammenfassung

Inclusion of Eigenvalues of General Eigenvalue Problems for Matrices. In this paper, a procedure for calculating bounds to eigenvalues of general matrix eigenvalue problems is proposed. The procedure is based on Temple quotients and their generalization by Lehmann, in connection with interval arithmetic. Numerical examples illustrate the fact that bounds for multiple or clustered eigenvalues can be calculated as well.

Einschließung von Eigenwerten bei allgemeinen Eigenwertaufgaben mit Matrizen. In dieser Arbeit wird ein Verfahren zur Berechnung von Schranken für Eigenwerte allgemeiner Matrixeigenwertaufgaben vorgeschlagen. Das Verfahren basiert auf Templeschen Quotienten und deren Verallgemeinerung von Lehmann in Verbindung mit Intervallarithmetik. Numerische Beispiele zeigen, daß auch Schranken für mehrfache oder dicht beieinander liegende Eigenwerte berechnet werden können.

AMS Subject Classification: primary: 65 F 15; secondary: 65 G 10

Keywords: bounds for eigenvalues; general matrix eigenvalue problems; interval arithmetic; Lehmann's method; Temple quotient; LDL^T decomposition

1. Introduction

Consider the general matrix eigenvalue problem

$$A x = \lambda B x, \quad A = A^T, \quad B = B^T, \quad B \text{ positive definite} \tag{1}$$

with real, generally nonsparse n by n matrices A and B, $n \in \mathbf{N}$. Let the eigenvalues be ordered by magnitude and let them be counted in accordance with their multiplicity: $\lambda_1 \leq \lambda_2 \leq \ldots \leq \lambda_n$.

A procedure for calculating sharp bounds for a f i x e d eigenvalue λ_j ($1 \leq j \leq n$) with the use of interval arithmetic is proposed. The procedure consists of three parts. In the *representative case* of an eigenvalue λ_j of *multiplicity one* with $1 < j < n$, these parts are:

1. Calculation of approximations $\tilde{\lambda}_{j-1}$, $\tilde{\lambda}_{j+1}$ and $\tilde{x}_j$ for λ_{j-1}, λ_{j+1} and for an eigenvector x_j belonging to the eigenvalue λ_j, with a suitable method [11]

2. Determination of a (rough) upper bound ρ for λ_{j-1} with $\lambda_{j-1} < \rho < \lambda_j$ and a (rough) lower bound σ for λ_{j+1} with $\lambda_j < \sigma < \lambda_{j+1}$ by means of the approximations $\tilde{\lambda}_{j-1}$ and $\tilde{\lambda}_{j+1}$ and of LDL^T decompositions [2]

3. Calculation of accurate bounds for λ_j by means of Temple quotients and their generalization by Lehmann using $\tilde{x}_j$, ρ and σ [7], [3]

The procedure can be applied for $j = 1$, $j = n$, and in the case of multiple eigenvalues as well. Its efficiency is illustrated by numerical examples.

At the end an eigenvalue problem with a differential equation is presented. Bounds for the smallest eigenvalues of this problem are calculated with the use of the proposed procedure.

2.   Bounds for Eigenvalues

For the purpose of calculating bounds to eigenvalues, Lehmann's method can be applied. The general Hilbert space version originates from [3] (see Appendix A); the following formulation is a special case with the view of treating matrix eigenvalue problems (1).

Lower Bounds

Theorem 1:

Let the following assumptions be valid:

1. A and B are real n by n matrices, $A = A^T$, $B = B^T$, B is positive definite.

2. Let $m \in N$; $u_1, u_2, \ldots, u_m$ are linearly independent elements of $\mathbb{R}^n$; $v_i := B^{-1} A u_i$ for $i = 1, \ldots, m$.

3. Let $\sigma \in \mathbb{R}$. The matrices A_0, A_1, A_2, $\hat{A}$, $\hat{B}$ are defined by

$$A_0 := \left(u_i^T B u_k \right)_{i,k=1,\ldots,m},$$

$$A_1 := \left(u_i^T A u_k \right)_{i,k=1,\ldots,m},$$

$$A_2 := \left(v_i^T A u_k \right)_{i,k=1,\ldots,m},$$

$$\hat{A} := A_1 - \sigma A_0 \text{ and } \hat{B} := A_2 - 2\sigma A_1 + \sigma^2 A_0.$$

$\hat{B}$ is positive definite. [1]

4. The eigenvalues μ_i of the matrix eigenvalue problem $\hat{A} x = \mu \hat{B} x$ are ordered by magnitude: $\mu_1 \leq \mu_2 \leq \ldots \leq \mu_m$.

Assertion: For all $l \in N$ with $l \leq m$ and $\mu_l < 0$, the interval $[\sigma + \frac{1}{\mu_l}, \sigma]$ contains at least l eigenvalues of the problem $A x = \lambda B x$.

The object of theorem 1 in this context is the calculation of bounds for eigenvalues of a matrix eigenvalue problem of dimension n by solving matrix eigenvalue problems of considerably smaller dimensions m.

---

[1] If σ is not an eigenvalue of the problem $A x = \lambda B x$, then $\hat{B}$ is always positive definite.

In order to use the theorem for the calculation of lower bounds for eigenvalues specified by their indices, additional information on the parameter σ is required: If σ is a rough lower bound for λ_p with $\lambda_{p-1} < \sigma < \lambda_p$, and if q is the number of negative eigenvalues of $\hat{A} x = \mu \hat{B} x$, then, with theorem 1, $\sigma + \frac{1}{\mu}$ is a lower bound for λ_{p-l} $(1 \le l \le q)$. These bounds are in general very accurate.

If $m = 1$ or if $\sigma > \lambda_n$, it is advantageous to use a corollary of theorem 1.

Corollary 1:

Let the assumptions 1 and 2 of theorem 1 be valid, and let σ, A_0, A_1 and A_2 be defined, as in 3. Let σ not be an eigenvalue of the problem $A x = \lambda B x$. Let the eigenvalues τ_i of the eigenvalue problem $(A_2 - \sigma A_1) x = \tau (A_1 - \sigma A_0) x$ be ordered by magnitude: $\tau_1 \le \tau_2 \le \ldots \le \tau_m$. If q is the number of eigenvalues τ_i smaller than σ, then the interval $[\tau_i, \sigma]$ contains at least $q + 1 - i$ eigenvalues of the problem $A x = \lambda B x$.

Proof: For a real number μ, $\mu \ne 0$, the following is valid:
μ is an eigenvalue of the problem $(A_1 - \sigma A_0) x = \tilde{\tau} (A_2 - 2\sigma A_1 + \sigma^2 A_0) x \iff$
$\frac{1}{\mu}$ is an eigenvalue of the problem $((A_2 - \sigma A_1) - \sigma (A_1 - \sigma A_0)) x =$

$$= \hat{\tau} (A_1 - \sigma A_0) x \iff$$

$\sigma + \frac{1}{\mu}$ is an eigenvalue of the problem $(A_2 - \sigma A_1) x = \tau (A_1 - \sigma A_0) x$.

Hence, the assertion follows from theorem 1, since $\hat{B}$ is positive definite.

As explained above, the parameter σ has to be determined appropriately, in order to calculate a lower bound for λ_j $(1 \le j \le n)$. Since the inclusion theorem again leads to a matrix eigenvalue problem (of dimension m), a determination of σ such that m is as small as possible is recommended. On the other hand, σ should *not* be placed in a cluster of eigenvalues.

$m = 1$ can be achieved, if $\lambda_j < \sigma < \lambda_{j+1}$ holds. In accordance with corollary 1, then, only the (Temple) quotient

$$\tau(\sigma) = \frac{A_2 - \sigma A_1}{A_1 - \sigma A_0}$$

must be calculated; one gets $\tau(\sigma) \le \lambda_j$.

If it should happen that this choice of σ is impossible — for instance, if λ_j is a multiple eigenvalue, or if the eigenvalues are clustered —, one can use σ such that $\lambda_s < \sigma < \lambda_{s+1}$ with $s \ge j$ and has to solve a matrix eigenvalue problem of dimension $m \ge s - j + 1$. Trying to separate clustered eigenvalues is not advantageous. The bounds are in general better, if σ is a (rough) lower bound for λ_{s+1}, where λ_s and λ_{s+1} are not clustered. Theorem 1 is applied instead of corollary 1 if $m > 1$, since the positive definiteness of $\hat{B}$ is useful for the further calculation.

If $\sigma > \lambda_n$, the passage to the limit $\sigma \to \infty$ can be carried out in the problem $(A_2 - \sigma A_1) x = \tau (A_1 - \sigma A_0) x$ in accordance with corollary 1. This results in the problem $A_1 x = \Lambda A_0 x$ with the positive definite matrix A_0; its eigenvalues Λ_k yield the following lower bounds: $\Lambda_k \le \lambda_{n-m+k}$ $(1 \le k \le m)$, that is $\Lambda_{m-(n-j)} \le \lambda_j$. (Incidentally, this special case is the well known procedure of Rayleigh–Ritz.)

Upper Bounds

The following theorem is used for the calculation of upper bounds.

Theorem 2:

Let the assumptions 1 and 2 from theorem 1 be valid. Let $\rho \in \mathbb{R}$, the matrices A_0, A_1, A_2 are defined as in 3, $\hat{A} := A_1 - \rho A_0$ and $\hat{B} := A_2 - 2\rho A_1 + \rho^2 A_0$. Let $\hat{B}$ be positive definite. For $i = 1, \ldots, m$, μ_i is the i-th largest eigenvalue of the matrix eigenvalue problem $\hat{A} x = \mu \hat{B} x$.

Assertion: For all $l \in N$ with $l \leq m$ and $\mu_l > 0$, the interval $[\rho, \rho + \frac{1}{\mu_l}]$ contains at least l eigenvalues of the problem $A x = \lambda B x$.

Proof: If one considers the problem $-A x = \lambda B x$ with the parameter $-\rho$, instead of the problem $A x = \lambda B x$ with the parameter σ, the proof is a straightforward consequence of theorem 1.

For the choice of ρ, points of view which are analogous to those for the choice of σ are valid. ρ has to separate two successive eigenvalues (i.e. $\lambda_{r-1} < \rho < \lambda_r$ with $r \leq j$), or has to be below λ_1.

If one has determined ρ and σ with $\rho < \lambda_r \leq \lambda_j \leq \lambda_s < \sigma$ (λ_r should be the smallest and λ_s the greatest eigenvalue of the cluster containing λ_j) and defines $m = s - r + 1$, the same elements u_i and thus the same matrices A_0, A_1, and A_2 can be used in theorem 1 and theorem 2. Of course, this yields bounds not only for λ_j, but for $\lambda_r, \ldots, \lambda_s$ as well.

LDL^T Decompositions

The next lemma [10] plays an important part in the determination of ρ and σ, as well as in the calculation of bounds for eigenvalues of matrices with small dimensions.

Lemma 1:

Let $\tilde{A}$, $\tilde{B}$ be real q by q matrices, $\tilde{A} = \tilde{A}^T$, $\tilde{B} = \tilde{B}^T$, let $\tilde{B}$ be positive definite, let $c \in \mathbb{R}$. The number of eigenvalues $\tilde{\lambda}_i$ of $\tilde{A} x = \tilde{\lambda} \tilde{B} x$ which are smaller than, equal to, or greater than c is equal to the number of negative, zero, or positive eigenvalues of the matrix $(\tilde{A} - c \tilde{B})$, respectively.

The number of negative, zero, and positive eigenvalues (the inertia) of a real symmetric matrix $\tilde{A}$ can be determined easily with the help of a decomposition $\tilde{A} = LDL^T$ [2]. Here, L is a regular lower triangular matrix, and D is a block diagonal matrix with blocks of order 1 or 2. $\tilde{A}$ and D possess the same inertia.

By starting with an approximation $\tilde{\lambda}_k$ for the k-th eigenvalue of $A x = \lambda B x$, a real number c is chosen such that $\lambda_{k-1} < c < \lambda_k$ or $\lambda_k < c < \lambda_{k+1}$ can be proved with an LDL^T decomposition of the matrix $(A - c B)$. In this way, the rough bounds ρ and σ are determined. The procedure is also well suited for calculating bounds to a certain eigenvalue by continued interval bisection for matrices of small dimensions.

The LDL^T decomposition should be considered as a subalgorithm of the proposed procedure; this subalgorithm can be changed if desired.

3. Numerical Application

The approximations $\tilde{\lambda}_{r-1}$, $\tilde{\lambda}_{s+1}$ (and $\tilde{x}_i$, $r \leq i \leq s$) required for the procedure can be calculated with one of the well known algorithms [11].

The LDL^T decomposition is implemented in analogy with [2] by using interval arithmetic. The strategy for the determination of pivot elements can be the same as in [2], if one uses, for example, the number with the greatest absolute value of the corresponding interval instead of the element itself for the comparisons. This yields an interval matrix $[D]$ with blocks of order 1 or 2 in the diagonal. If an interval with zero included should occur during the determination of the inertia for $[D]$ — that is, if zero is included in a block of order 1 or in the determinant of a block of order 2 —, it is impossible to decide whether this interval corresponds to a negative, zero, or positive eigenvalue. The shift parameter c is then altered slightly, and the decomposition is repeated.

For the application of theorems 1 and 2, the manner of choosing the vectors $u_1, \ldots, u_m$ must be explained. Let $\rho < \lambda_r \leq \lambda_j \leq \lambda_s < \sigma$, $m = s - r + 1$, and let $x_r, \ldots, x_s$ be eigenvectors which are associated with $\lambda_r, \ldots, \lambda_s$. Then, the approximations $\tilde{x}_r, \ldots, \tilde{x}_s$ are used for $u_1, \ldots, u_m$, because of results from [5] dealing with the convergence of Lehmann's bounds. (For $s = 1$ these results say, that "good" approximations for the eigenvectors yield "good" lower bounds. Furthermore, it is easy to see that corollary 1 results in the exact eigenvalues of problem (1), if $u_1, \ldots, u_m$ are the exact eigenvectors.) With subspace iteration, these approximations can be improved, if necessary. Then — if B is not the identity — guaranteed error bounds for the solutions v_i of the linear equations $B v_i = A u_i$ ($i = 1, \ldots, m$) are required. For this purpose, methods based on the fixed point theorem of Brouwer can be applied [9].

It should be mentioned here that this need can be overcome. Goerisch's generalization [3] of Lehmann's method allows the use of approximations $\tilde{v}_i$, instead of guaranteed error bounds for the exact solutions v_i. — An extension to the sparse matrix technique is possible.

Theorems 1 and 2 result in a matrix eigenvalue problem $[\hat{A}] x = \mu [\hat{B}] x$ with interval matrices $[\hat{A}]$ and $[\hat{B}]$; this problem is solved with a bisection method based on the LDL^T decomposition, if $m > 1$. Since $u_1, \ldots, u_m$ are approximations for the eigevectors, $[\hat{A}]$ and $[\hat{B}]$ are almost diagonal matrices, that is, the off-diagonal elements are relatively small. Consequently, the LDL^T decomposition is applied to a matrix $[\hat{A}] - c [\hat{B}]$ of the same structure.

In practice, the use of a spectral shifted matrix eigenvalue problem $(A - \gamma_j B) x = (\lambda - \gamma_j) B x$ instead of problem (1) turns out to be advantageous. The shift γ_j is determined such that $\lambda_j - \gamma_j$ is close to zero.

4. Numerical Examples

All numerical results given below have been calculated in PASCAL–SC [6] with interval arithmetic using 13 decimal places; in case of the LDL^T decomposition, interval arithmetic with 21 decimal places has been used.

The first example [11] has also been treated by Alefeld [1]. Let

$$
F := \begin{pmatrix} 10 & 2 & 3 & 1 & 1 \\ 2 & 12 & 1 & 2 & 1 \\ 3 & 1 & 11 & 1 & -1 \\ 1 & 2 & 1 & 9 & 1 \\ 1 & 1 & -1 & 1 & 15 \end{pmatrix} \text{ and } G := \begin{pmatrix} 12 & 1 & -1 & 2 & 1 \\ 1 & 14 & 1 & -1 & 1 \\ -1 & 1 & 16 & -1 & 1 \\ 2 & -1 & -1 & 12 & -1 \\ 1 & 1 & 1 & -1 & 11 \end{pmatrix}.
$$

In the case of the problem $F x = \lambda G x$, all eigenvalues can be separated with the help of the LDL^T decomposition. The decompositions result in

$$
\begin{aligned}
\lambda_1 &\in [4.3^{33}_{23} \ E - 1] \,, \\
\lambda_2 &\in [6.6^{5}_{3} \ E - 1] \,, \\
\lambda_3 &\in [9.4^{5}_{2} \ E - 1] \,, \\
\lambda_4 &\in [1.1^{11}_{08} \ E + 0] \,, \\
\lambda_5 &\in [1.49^{4}_{0} \ E + 0] \,;
\end{aligned}
$$

therefore, one can always calculate with $m = 1$. The accurate inclusion intervals are

$$
\begin{aligned}
\lambda_1 &\in [4.327\ 872\ 110\ 1^{70}_{69} \ E - 1] \,, \\
\lambda_2 &\in [6.636\ 627\ 483\ 92^{4}_{3} \ E - 1] \,, \\
\lambda_3 &\in [9.438\ 590\ 046\ 68^{4}_{3} \ E - 1] \,, \\
\lambda_4 &\in [1.109\ 284\ 540\ 01^{8}_{7} \ E + 0] \,, \\
\lambda_5 &\in [1.492\ 353\ 232\ 54^{4}_{2} \ E + 0] \,.
\end{aligned}
$$

In the case of the problem $G x = \lambda F x$, one can again always calculate with $m = 1$ and obtains:

$$
\begin{aligned}
\lambda_1 &\in [6.700\ 826\ 441\ 04^{3}_{2} \ E - 1] \,, \\
\lambda_2 &\in [9.014\ 819\ 587\ 98^{7}_{6} \ E - 1] \,, \\
\lambda_3 &\in [1.059\ 480\ 277\ 30^{2}_{1} \ E + 0] \,, \\
\lambda_4 &\in [1.506\ 789\ 408\ 3^{60}_{59} \ E + 0] \,, \\
\lambda_5 &\in [2.310\ 604\ 321\ 34^{9}_{8} \ E + 0] \,.
\end{aligned}
$$

Let the second example [8, p. 123] be the problem $A x = \lambda x$ with

$$
A := \begin{pmatrix} M & C_2^T & & 0 \\ C_2 & M & C_3^T & \\ & C_3 & M & C_4^T \\ 0 & & C_4 & M \end{pmatrix}, \quad C_i := (e_6 * e_1^T - e_1 * e_6^T)/10^{2i} \text{ for } i = 2, 3, 4,
$$

$$
M := \begin{pmatrix} J & K^T \\ K & J \end{pmatrix}, \quad J := \begin{pmatrix} 1 & 2 & 3 \\ 2 & 4 & 5 \\ 3 & 5 & 6 \end{pmatrix},
$$

$$
K := \begin{pmatrix} 0 & -1 & -0.5 \\ 1 & 0 & -0.333\ 333\ 3 \\ 0.5 & 0.333\ 333\ 3 & 0 \end{pmatrix}.
$$

The matrix M has three double eigenvalues; therefore, the eigenvalues of A are arranged in three clusters, each containing eight eigenvalues. Here the effect of trying to separate clustered eigenvalues with the LDL^T decomposition can be illustrated. If, for example, the eigenvalues 17 to 24 are considered, the approximations are

$$\tilde{\lambda}_{17} = 1.143\ 482\ 563\ 420\ E + 1\ ,$$
$$\tilde{\lambda}_{18} = 1.143\ 482\ 563\ 426\ E + 1\ ,$$
$$\tilde{\lambda}_{19} = 1.143\ 484\ 925\ 579\ E + 1\ ,$$
$$\tilde{\lambda}_{20} = 1.143\ 484\ 925\ 579\ E + 1\ ,$$
$$\tilde{\lambda}_{21} = 1.143\ 484\ 926\ 049\ E + 1\ ,$$
$$\tilde{\lambda}_{22} = 1.143\ 484\ 926\ 049\ E + 1\ ,$$
$$\tilde{\lambda}_{23} = 1.143\ 487\ 288\ 255\ E + 1\ ,$$
$$\tilde{\lambda}_{24} = 1.143\ 487\ 288\ 255\ E + 1\ .$$

Without considering the structure of A, it can be expected that the corresponding eigenvalues are clustered. If, nevertheless, one tries to separate the eigenvalues with LDL^T decompositions and halves the intervals if necessary, this results in the following inclusions:

$$\lambda_{17,18} \in [1.143\ 48^{44}_{07} \qquad E+1]\ ,$$
$$\lambda_{19,20} \in [1.143\ 484\ 925^{9}_{4} \quad E+1]\ ,$$
$$\lambda_{21,22} \in [1.143\ 484\ 926\ 0^{6}_{5}\ E+1]\ ,$$
$$\lambda_{23,24} \in [1.143\ 487\ 2^{90}_{87} \qquad E+1]\ .$$

(Starting with λ_{17}, the calculation is stopped as soon as successive eigenvalues are separated; the inclusions could be much more accurate, but it is expensive.) For the purpose of calculating better inclusions for λ_{19}, one can use $\varsigma = 1.143\ 484\ 4\ E + 1$ (an upper bound for λ_{18}) and $\sigma = 1.143\ 484\ 926\ 05\ E + 1$ (a lower bound for λ_{21}). With $m = 21 - 18 - 1 = 2$, the same 2 by 2 matrices A_0, A_1 and A_2 can be used in theorems 1 and 2. The calculated inclusion is

$$\lambda_{19} \in [1.143\ 484\ 92^{6}_{3}E + 1]\ ;$$

the result is worse than that obtained by the LDL^T decompositions. If one does not try to separate clustered eigenvalues, only one decomposition has to be calculated for each rough bound; this yields

$$\lambda_i \in [-\ 8.^{6}_{9}\ E - 1] \quad \text{for } i = 1,\ldots,8\ ,$$
$$\lambda_i \in [\quad 4.^{48}_{38}\ E - 1] \quad \text{for } i = 9,\ldots,16\ ,$$
$$\lambda_i \in [\quad 1.1^{6}_{3}\ E + 1] \quad \text{for } i = 17,\ldots,24\ .$$

(The lower bound for λ_1 and the upper bound for λ_{24} are not used.) The smallest possible m in this situation is $m = 8$. The inclusion intervals obtained are

$$\lambda_{1,2} \quad \in [-8.777\ 193\ 981\ 43^{6}_{8}\ E - 1]\ ,$$
$$\lambda_{3,4} \quad \in [-8.776\ 888\ 529\ 5^{29}_{31}\ E - 1]\ ,$$
$$\lambda_{5,6} \quad \in [-8.776\ 888\ 468\ 44^{2}_{3}\ E - 1]\ ,$$
$$\lambda_{7,8} \quad \in [-8.776\ 583\ 042\ 16^{3}_{4}\ E - 1]\ ,$$

$$\lambda_{9,10} \quad \in [\ 4.428\ 326\ 699\ 87^{7}_{6}\ E - 1]\ ,$$
$$\lambda_{11,12} \in [\ 4.428\ 395\ 910\ 60^{7}_{5}\ E - 1]\ ,$$
$$\lambda_{13,14} \in [\ 4.428\ 395\ 924\ 45^{1}_{0}\ E - 1]\ ,$$
$$\lambda_{15,16} \in [\ 4.428\ 465\ 156\ 18^{8}_{7}\ E - 1]\ ,$$

$$\lambda_{17,18} \in [\ 1.143\ 482\ 563\ 42^{3}_{2}\ E + 1]\ ,$$
$$\lambda_{19,20} \in [\ 1.143\ 484\ 925\ 57^{9}_{8}\ E + 1]\ ,$$
$$\lambda_{21,22} \in [\ 1.143\ 484\ 926\ 05^{1}_{0}\ E + 1]\ ,$$
$$\lambda_{23,24} \in [\ 1.143\ 487\ 288\ 25^{3}_{2}\ E + 1]\ .$$

Appendix

A Lehmann's Method

The Hilbert space version of Lehmann's method is taken from [3].

Theorem 3:

Let the following assumptions and notations be valid:

1. $(H, \langle . | . \rangle)$ is a real Hilbert space.

2. $D(M)$, $D(N)$, $D(S)$ are real vector spaces with $D(M) \subset D(N) \subset D(S)$.

3. $M : D(M) \to H$, $N : D(N) \to H$, $S : D(S) \to H$ are linear mappings.

4. $\langle S f | N g \rangle = \langle S g | N f \rangle$ holds true for all $f, g \in D(N)$;
   $\langle S f | N f \rangle > 0$ holds true for all $f \in D(N)$ with $f \neq 0$.

5. $\langle S f | M g \rangle = \langle S g | M f \rangle$ holds true for all $f, g \in D(N^{-1}M)$ [2].

6. There exist a sequence (λ_i) of eigenvalues and a sequence (ϕ_i) of eigenelements of the problem $M \phi = \lambda N \phi$ such that

$$M \phi_i = \lambda_i N \phi_i,$$

$$\langle S \phi_i | N \phi_k \rangle = \delta_{i,k} \quad (\delta_{i,k} \text{ Kronecker's symbol) and}$$

$$\langle S f | N f \rangle = \sum_i \langle S f | N \phi_i \rangle^2 \quad \text{for all } f \in D(N^{-1}M).$$

7. Let $m \in N$; $u_1, u_2, \ldots, u_m$ are linearly independent elements of $D(N^{-1}M)$;
   $v_i := N^{-1}M u_i$ for $i = 1, \ldots, m$.

8. Let $\sigma \in \mathbb{R}$. The matrices A_0, A_1, A_2, $\hat{A}$, $\hat{B}$ are defined by

$$A_0 := \left(\langle S u_i | N u_k \rangle \right)_{i,k=1,\ldots,m},$$

$$A_1 := \left(\langle S u_i | M u_k \rangle \right)_{i,k=1,\ldots,m},$$

$$A_2 := \left(\langle S v_i | M u_k \rangle \right)_{i,k=1,\ldots,m},$$

$$\hat{A} := A_1 - \sigma A_0 \text{ and } \hat{B} := A_2 - 2 \sigma A_1 + \sigma^2 A_0.$$

$\hat{B}$ is positive definite.[3]

9. For $i = 1, \ldots, m$, μ_i is the i-th smallest eigenvalue of the matrix eigenvalue problem $\hat{A} x = \mu \hat{B} x$.

Assertion: For all $l \in N$ with $l \leq m$ and $\mu_l < 0$, the interval $[\sigma + \frac{1}{\mu_l}, \sigma]$ contains at least l eigenvalues of the problem $M \phi = \lambda N \phi$.

---

[2] If P is a linear mapping, $D(P)$ denotes its domain of definition.
[3] If σ is not an eigenvalue of the problem $M \phi = \lambda N \phi$, then $\hat{B}$ is always positive definite.

By the following definitions, theorem 1 follows from this theorem:

1. $H := \mathbb{R}^n$, $\langle x|y \rangle := \sum_{i=1}^{n} x_i y_i$ for $x, y \in \mathbb{R}^n$.

2. $D(M) := \mathbb{R}^n$, $D(N) := \mathbb{R}^n$, $D(S) := \mathbb{R}^n$.

3. $M : \mathbb{R}^n \ni x \mapsto M x := A x \in \mathbb{R}^n$, $N : \mathbb{R}^n \ni x \mapsto N x := B x \in \mathbb{R}^n$ and
   $S : \mathbb{R}^n \ni x \mapsto S x := x \in \mathbb{R}^n$.

Assumptions 4, 5, and 6 are fulfilled because of the properties of A and B.

B   Application to an Eigenvalue Problem with a Differential Equation

For the purpose of calculating bounds to eigenvalues of an eigenvalue problem with a differential equation, theorem 3 can be used. This results in a matrix eigenvalue problem of the form (1); hence, the proposed procedure can be used to solve it.

Let the Mathieu differential equation

$$-f''(x) + 2 h^2 \cos(2x) f(x) = \lambda f(x) ,$$

$$\left. \begin{array}{l} f(x) = -f(-x) \\ f(x) = f(x + \pi) \end{array} \right\} \text{ for } x \in \mathbb{R} ,$$

$$h \in \mathbb{R}$$

be given. The problem fits into theorem 1 as follows:

1. $H := \{ f \in L_2(\mathbb{R}) \mid f(x) = -f(-x), \ f(x) = f(x + \pi) \text{ for } x \in \mathbb{R} \}$,
   $\langle f|g \rangle := \int_0^\pi f(x) \, g(x) \, dx$ for $f, g \in H$.

2. $D(M) := H \cap C^2(\mathbb{R})$, $D(N) := H$, $D(S) := H$.

3. The operators M, N, and S are defined by $M f(x) := -f''(x) + 2 h^2 \cos(2x) f(x)$
   for
   $f \in D(M)$, $x \in \mathbb{R}$, $N f(x) := f(x)$ for $f \in D(N)$, $x \in \mathbb{R}$, $S f(x) := f(x)$ for
   $f \in D(S)$, $x \in \mathbb{R}$.

Thus, assumptions 4, 5, and 6 are fulfilled.

7. $u_k(x) := \sin(2 k x)$ for $x \in \mathbb{R}$, $k \in \mathbb{N}$.

8. $4p^2 - 2h^2$ is a (rough) lower bound for the p-th smallest eigenvalue λ_p of the
   problem $M \phi = \lambda N \phi$. The scalar products result in
   $\langle S u_i|N u_k \rangle = \frac{1}{2} \pi \delta_{i,k}$,
   $\langle S u_i|M u_k \rangle = 2 \pi k^2 \delta_{i,k} + \frac{1}{2} \pi h^2 \delta_{k \pm 1, i}$,
   $\langle S v_i|M u_k \rangle = 8 \pi k^4 \delta_{i,k} + 2 \pi h^2 (i^2 + k^2) \, \delta_{k \pm 1, i} + \frac{1}{2} \pi h^4 (-\delta_{i,2-k} + 2 \delta_{i,k} + \delta_{k \pm 2, i})$.

Hence the matrix eigenvalue problem $\hat{A} x = \mu \hat{B} x$ can be established. If it is solved with the proposed procedure, guaranteed lower bounds for eigenvalues of the Mathieu differential equation can be derived from the bounds for the eigenvalues μ. The procedure of Rayleigh–Ritz for calculating upper bounds results in the matrix eigenvalue problem $A_1 x = \Lambda A_0 x$. At the same time, an upper bound for the i-th eigenvalue Λ_i is an upper bound for λ_i. Bounds for the smallest eigenvalues of the Mathieu differential equation (calculated with matrices of dimension 15) are listed in the following table.

$h^2 = 1$	$h^2 = 7$
$\lambda_1 \in [3.917\ 024\ 772\ 99^9_5\ E + 0]$	$\lambda_1 \in [5.175\ 454\ 066\ ^{200}_{196}\ E - 1]$
$\lambda_2 \in [1.603\ 297\ 008\ 14^1_0\ E + 1]$	$\lambda_2 \in [1.702\ 666\ 078\ 32^8_7\ E + 1]$
$\lambda_3 \in [3.601\ 428\ 991\ 06^3_2\ E + 1]$	$\lambda_3 \in [3.670\ 350\ 271\ 26^8_5\ E + 1]$
$\lambda_4 \in [6.400\ 793\ 718\ 92^5_4\ E + 1]$	$\lambda_4 \in [6.439\ 053\ 417\ 6^{72}_{68}\ E + 1]$

Lehmann's inclusion theorem for problems with positive definite operators on the left, Goerisch's generalizations of Lehmann's methods [3], [4], as well as numerous other methods result in matrix eigenvalue problems of type (1). Hence, the proposed procedure can frequently be used in connection with the calculation of bounds for eigenvalues of eigenvalue problems with differential equations.

References

[1] Alefeld, G.: Berechenbare Fehlerschranken für ein Eigenpaar beim verallgemeinerten Eigenwertproblem. To appear

[2] Bunch, J.R.; Kaufmann, L.; Parlett, B.N.: Decomposition of a Symmetric Matrix. Numer. Math. 27 (1976), 95–109

[3] Goerisch, F.; Albrecht, J.: Eine einheitliche Herleitung von Einschließungssätzen für Eigenwerte. In: J. Albrecht, L. Collatz und W. Velte (Hrsg.): Numerische Behandlung von Eigenwertaufgaben, Band 3. International Series of Numerical Mathematics (ISNM), Vol 69. Basel–Boston–Stuttgart: Birkhäuser 1984, 58–88

[4] Goerisch, F.; Haunhorst, H.: Eigenwertschranken für Eigenwertaufgaben mit partiellen Differentialgleichungen. Z. Angew. Math. Mech. 65 (1985), 129–135

[5] Knyazev, A. V.: Convergence Rate Estimates for Iterative Methods for Mesh Symmetric Eigenvalue Problem. Sov. J. Numer. Anal. Modelling, V. 2, N. 5 (1987), 371–396

[6] Kulisch, U.: PASCAL–SC: A PASCAL Extension for Scientific Computation; Information Manual and Floppy Disks. Stuttgart: Teubner; Chichester, New York, Brisbane, Toronto, Singapore: Wiley 1987

[7] Lehmann, N.J.: Beiträge zur Lösung linearer Eigenwertprobleme I, II. Z. Angew. Math. Mech. 29 (1949), 341–356; 30 (1950), 1–16

[8] Parlett, B.N.: The Symmetric Eigenvalue Problem. Englewood Cliffs: Prentice Hall 1980

[9] Rump, S.M.: Solution of Linear and Nonlinear Algebraic Problems with Sharp, Guaranteed Bounds. Computing Suppl. 5 (1984), 147–168

[10] Schwarz, H.R.: Methode der finiten Elemente. Stuttgart: Teubner 1980

[11] Wilkinson, J.H.; Reinsch, C.: Handbook for Automatic Computation. Volume II. Linear Algebra. Berlin: Springer 1971

Dipl.-Math. Henning Behnke
Institut für Mathematik
Technische Universität Clausthal
Erzstrasse 1
D-3392 Clausthal-Zellerfeld
Federal Republic of Germany

Computing, Suppl. 6, 79–88 (1988)

Verified Inclusion for Eigenvalues of Certain Difference and Differential Equations

Martin Ohsmann, Aachen

Abstract — Zusammenfassung

Verified Inclusion for Eigenvalues of Certain Difference and Differential Equations. The paper presents a method for obtaining accurate verified inclusions of eigenvalues. Second order difference and differential equations with a boundary condition at infinity can be treated. Using interval arithmetic and the exact scalar product very accurate inclusions can be obtained even for ill conditioned problems where classical methods fail.

Verifizierter Einschluß von Eigenwerten gewisser Differenzen- und Differentialgleichungen. Es wird ein einfaches Verfahren zum hochgenauen verifizierten Einschluß von Eigenwerten angegeben. Es werden Randwertprobleme von Differenzen- und Differentialgleichungen 2. Ordnung mit einer Randbedingung im Unendlichen behandelt. Es wird gezeigt, daß durch die Verwendung der Intervall-arithmetik und des exakten Skalarproduktes auch bei sehr schlecht konditionierten Problemen eine hohe Genauigkeit erreicht werden kann.

1 Introduction

In this paper a method is described for obtaining verified inclusions of real eigenvalues of a certain type of difference equations. The method has a simple intuitive interpretation and is easy to implement, yet it provides high accuracy. An ill-conditioned example will demonstrate the advantages of this procedure. Furthermore the idea can be carried over to the problem of computing eigenvalues of certain second order ordinary differential equations on the semiinfinite interval.

The method will be worked out using a simple example. Consider the well known Mathieu equation :

$$y''(x) + (\lambda - 2q\cos(2x))y(x) = 0 \tag{1}$$

Here q denotes a given parameter and the eigenvalue problem may be stated as follows :

$$\text{Find real } \lambda \text{ , such that } y(x) \text{ in (1) has period } \pi \text{ and is even.} \tag{2}$$

Following Wimp [1] the problem may be treated using a Fourier series development.

$$y(x) = \frac{1}{2}y_0 + \sum_{k=1}^{\infty} y_k \cos(2kx) \tag{3}$$

Inserting (3) into (1) leads to the following difference equation for the Fourier coefficients y_k:

$$\frac{\lambda}{2}y_0 - qy_1 \;=\; 0 \tag{4}$$

$$y_k - \frac{\lambda - 4(k+1)^2}{q}y_{k+1} + y_{k+2} \;=\; 0 \qquad k = 0, 1, \ldots \tag{5}$$

Clearly (4,5) has a solution for each λ, but only certain values of λ will lead to solutions of the original eigenvalue problem (1). To yield a solution of (1) the series (3) has to converge. To meet at least a necessary condition for convergence it is therefore meaningful to pose the following difference equation eigenvalue problem

$$\text{Find real} \quad \lambda \quad \text{such that} \quad \lim_{k\to\infty} y_k(\lambda) = 0 \quad \text{in (4,5)} . \tag{6}$$

At this point it is not clear that this necessary condition is also sufficient for obtaining a solution of (2). This question will be considered later. The equations (4,5) together with the condition (6) present an example of a difference equation eigenvalue problem. A method for obtaining verified bounds for eigenvalues will be presented in the following.

2 Inclusion of Difference Equation Eigenvalues

2.1 The Method

The problems to be solved are of the following form:

Assume $f_0(\lambda)$, $f_1(\lambda)$, $a_{k+1}(\lambda)$ and $b_k(\lambda)$ to be real continous functions in an interval $I \subset \mathbb{R}$ and $a_{k+1}(\lambda)$ having no zeroes in I. $(k = 0, 1, \ldots)$.

Now consider the difference equation

$$x_k = b_k x_{k+1} + a_{k+1} x_{k+2} \tag{7}$$

together with the initial values

$$x_0 = f_0(\lambda) \quad , \quad x_1 = f_1(\lambda) \tag{8}$$

and the eigenvalue condition

$$\lim_{k\to\infty} x_k(\lambda) = 0 \tag{9}$$

The method to obtain inclusions for the real (simple) eigenvalues of this problem is based on the following observation.

Computing the solution of (7,8) for some values of λ one notices the following behaviour of $x_k(\lambda)$ (see Fig. 1): At the beginnig the values x_k oscillate. After a while a monotonic behaviour settles and the values seem to diverge monotonically. Whether they tend to $+\infty$ or $-\infty$ depends on the value of λ.

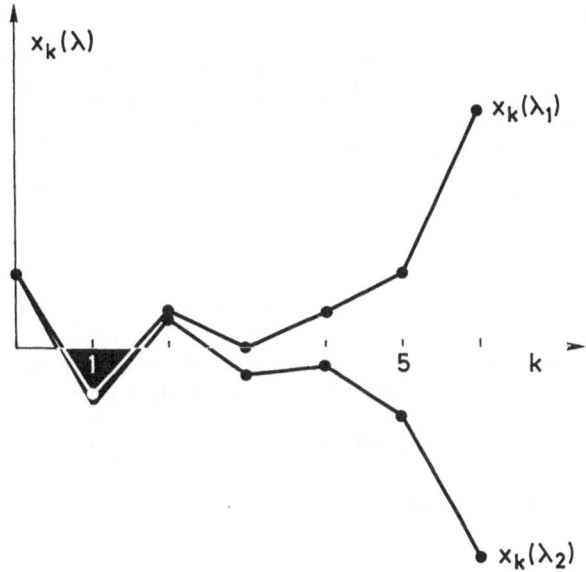

Figure 1: Typical sequences $x_k(\lambda)$ for values λ_1, λ_2 enclosing an eigenvalue.

This leads to the following conjecture:

If $x_k \to +\infty$ for λ_1 and $x_k \to -\infty$ for λ_2 then an eigenvalue λ will lie between λ_1 and λ_2.

Under certain assumptions concerning the growth of a_k and b_k this can be proved. This leads to a simple procedure for computing bounds for the desired eigenvalues. Starting with trial values one searches for λ_1, λ_2 exhibiting the desired behaviour. These two values give initial bounds for an eigenvalue of the difference equation. To get sharper bounds one simply bisects the interval $[\lambda_1, \lambda_2]$.

To get verified bounds all computations have to be done exactly or using interval arithmetic. Clearly, in practice only interval arithmetic provides a useful tool. Further it must be possible to check the statement " $x_k \to +\infty$ for λ_1 " on a computer .

As will be seen later it is sufficient to find m and k_0 such that $x_k(\lambda_1) \geq m > 0$ and $x_k(\lambda_2) \leq -m$ for $k \geq k_0$ to prove that there lies an eigenvalue between λ_1 and λ_2. A simple tool for deciding this fact after computing only a finite number of values x_k will be called a stopping rule. A simple stopping rule is the following :

Assume that there exists a number k_0 (depending on the interval I) such that

$$|b_k| - |a_{k+1}| > 1 \quad \text{and} \quad -\frac{b_k}{a_{k+1}} > 0 \quad \text{for} \quad k \geq k_0 \tag{10}$$

Supposed $x_{k+1} > x_k > m > 0$ for $k = k_0$ evaluation of (7) yields :

$$x_{k+2} = \frac{1}{a_{k+1}}(-b_k x_{k+1} + x_k) \tag{11}$$

$$> \frac{|-b_k|x_{k+1} - x_{k+1}}{|a_{k+1}|} > x_{k+1} \tag{12}$$

Thus by induction $x_k > m$ for $k \geq k_0$.

To compute the "sign" of the sequence x_k for a value λ_1 one proceeds as follows. Compute the sequence x_k up to the index k_0 using the difference equation and continue computing until you reach an index n where $x_{n+1} > x_n > 0$ or $x_{n+1} < x_n < 0$ then stop computing. The sign of the sequence then is the sign of x_n for this particular λ. The conjecture that an eigenvalue lies between two values λ_1, λ_2 with different "signs" can be justified by the proof sketched below.

2.2   Sketch of Proof

To prove the conjecture mentioned above the following assumptions will be made: There exist constants k_1 and $Q > 1$ independent of $\lambda \in I$ such that

$$|b_k| - |a_k| > max(|a_k|, Q) \quad \text{for all } k \geq k_1 \tag{13}$$

Now consider the following continued fraction approximant :

$$b_0 + \frac{a_1|}{|b_1} + \frac{a_2|}{|b_2} + \frac{a_3|}{|b_3} + \cdots + \frac{a_n|}{|b_n} = \frac{A_n}{B_n} \tag{14}$$

(Notation of Perron [4]) . According to Perron, one has the following representation of x_k :

$$x_{k+2} = u_k \quad \left(\ f_0(\lambda) - \frac{A_k(\lambda)}{B_k(\lambda)} f_1(\lambda) \ \right) \tag{15}$$

where u_k is defined as :

$$u_k = \frac{B_k}{(-1)^k a_1 a_2 \cdots a_{k+1}} \tag{16}$$

This representation of x_k gives insight into the behaviour of x_k for large k. For large k one can prove that u_k does not change its sign. Thus the sign of x_k is entirely determined by the bracket in (15).

Under the assumptions made above the continued fraction A_k/B_k converges to a continuous function $C(\lambda)$ of $\lambda \in I$. If therefore x_k exhibit a sign change for different values of λ this can only be due to a sign change of the expression $f_0(\lambda) - C(\lambda)f_1(\lambda)$. Due to the fact that this expression is continuous with respect to λ it has a zero between the values λ_1 and λ_2.

It remains to show that the sequence $x_k(\lambda)$ associated with this particular value of λ tends to zero and therefore is an eigenvalue. This is the case due to the fact that u_k can only grow so slow, that its growing is dominated by the convergence of the bracket in (15) to zero. One even can prove the existence of a constant S such that

$$|x_k(\lambda)| < S(\frac{1}{Q})^k \quad \text{for} \quad k > k_1 \tag{17}$$

for this sequence. In the case of the example of the Mathieu equation this ensures the convergence of the Fourier series. The constant S can also be computed on a computer using interval arithmetic. This faciliates an eigenfunction inclusion.

2.3   Implementation

For the computation of eigenvalue bounds the difference equation (7) has to be evaluated using interval arithmetic. One may notice that the formula has the form of a scalar product. Evaluation of the recurrence relation using naive interval arithmetic leads to rapidly growing intervals containing the numbers x_k. For well conditioned problems experience has shown that even this blowing up is no major drawback, for in order to use the stopping rule one only has to decide whether $x_{k+1} > x_k > 0$. This decision can be performed with very bad inclusions of the values x_k.

If the eigenvalue interval $[\lambda_1, \lambda_2]$ gets very narrow after some bisection steps the following critical situation may arise: In evaluating the recurrence for λ_1 the intervals for x_k and x_{k+1} both may contain zero for some index k. This indicates that it is no longer possible to decide the sign of the sequence. Thus the blowing up was too large and the inclusion using naive interval arithmetic can not be narrowed. In well conditioned examples, as experience has shown, this happens when λ_1 and λ_2 aggree in about 12 digits when performing 15 digit computation. Thus one normally will be satisfied with this result.

If the user is not satisfied with this inclusion the evaluation of the recurrence can be performed using residual correction with the exact scalar product. One represents the values x_k using staggered corrections in this case. The number of staggered corrections to be used can be controlled automatically: Each time you encounter a critical situation a further staggered correction has to be used. Using this approach it was possible to achieve maximum digit accuracy of the eigenvalue bounds in all examples that have been tried.

It should further be noted, that the index n up to which the sequence x_k has to be computed usually seems to be small (less than 30) and grows large only for ill conditioned problems.

2.4   Description of Other Methods

Wimp[1] solves the Mathieu example given above using the Miller algorithm for the computation of $C(\lambda)$ in $f_0(\lambda) - C(\lambda)f_1(\lambda)$. To use the Miller algorithm one has to determine an a-priori index N of the continued fraction approximant to be used. Recomputations with different N have to be done. The computed zero λ depends on this index. Using larger and larger N one hopes that the zero stabilizes and will stop enlarging N at this point. No verification is obtained.

Another method is to treat the problem as a matrix eigenvalue problem involving an infinite matrix. One simply cuts this matrix at a reasonable (?) size N and computes the eigenvalues of this matrix and hopes that some of them (usually the small ones) agree with the difference equation eigenvalues. Different sizes of N are tried to confirm this suggestion . This method may lead to bad results as will be shown later and no verification is obtained.

2.5 Numerical Results

For $q = 60$ Wimp [1] gives the following results for the computation of an eigenvalue (characteristic exponent) of the Mathieu equation using the Miller algorithm with index N:

N	λ
10	264.38823665893
15	263.19071067355
20	263.19071067037
25	263.19071067037

Using the verified bisection method one obtains:

$$\lambda \in [263.190710670497765, 263.190710670497992]$$

Because the example is well conditioned the values of Wimp are very accurate. However one must notice that the computations have to be done for different N until convergence of the eigenvalue seems to have happened. Even with this extra amount of computation no verification is obtained.

As next example consider the following eigenvalue problem :

$$qy''(x) + (y(x)\sin(x))' = (\lambda/4)y(x) \quad y(x) \quad \text{even and } 4\pi \text{ periodic} \qquad (18)$$

where the eigenvalue next to zero has to be computed. This may be treated using a Fourier series as shown above. The resulting difference equation eigenvalue problem has been solved using the matrix method. The eigenvalues have been computed using the EISPACK CG Package on a Cyber 175. The following results have been obtained for $q = 1/32$:

N	λ
10	$-9.9...E - 2$
20	$-7.5...E - 6$
30	$-1.7...E - 11$
40	$-4.2...E - 11$
60	$-4.2...E - 11$
80	$-1.1...E - 10$

(N : size of the matrix , λ : smallest eigenvalue obtained)

The result obtained using the verified inclusion method presented above yields :

$$\lambda = -4.0517768.....E - 28$$

Using 2 staggered corrections with 15 decimal digit arithmetic the recurrence relation had to be computed up to $k = 64$ to obtain a maximum accuracy result.

The values obtained by the matrix method using conventional arithmetic are completely wrong. No stabilization of the eigenvalue can be observed and even the order of magnitude is wrong. Using larger matrices even makes things worse in this case because larger and larger eigenvalues of the matrix will get involved and make the problem more and more illconditioned .

3   Extensions

3.1   Transcendental Equations

The method presented solves equations of the form

$$f_0(\lambda) - C(\lambda)f_1(\lambda) = 0 \qquad (19)$$

where $C(\lambda)$ is defined as the limit of a continued fraction development. Some other problems therefore can be solved using this method including the following examples :

1) Evaluation of continued fractions :

Choose $f_0(\lambda) = \lambda$ and $f_1(\lambda) = 1$ and let C be independent of λ the continued fraction to be computed. Clearly any inclusion of the solution of (19) yields an inclusion for the value of the continued fraction. The method can be carried over to complex continued fractions in this case.

2) Computing zeroes of Bessel functions

Choose $f_0(\lambda) = 0$ and $f_1(\lambda) = 1$ and choose $C(\lambda) = J_0(\lambda)/J_1(\lambda)$ (the corresponding continued fraction can be found in almost all references to Bessel functions). Then each inclusion for a solution of (19) gives an inclusion for a zero of $J_0(\lambda)$. The method clearly applies to many other functions.

The assumptions made in order to justify the method have to be verified for each example seperately. Usually this is quite easy. In some cases it may be necessary to apply equivalence transformations of the involved continued fraction to fulfil the requirements of the algorithm.

3.2   Treatment of Differential Equations

To demonstrate how the ideas of the preceding section may be carried over to allow a treatment of certain 2nd order ordinary differential equation on the interval $[0, \infty)$ consider the following differential equation :

$$y''(x) = (f(x) - \lambda)y(x) \qquad x > 0 \qquad (20)$$

together with the initial conditions :

$$y(0) = g(\lambda) \qquad y'(0) = h(\lambda) \qquad (21)$$

and the boundary condition

$$\lim_{x \to \infty} y(x, \lambda) = 0 \qquad (22)$$

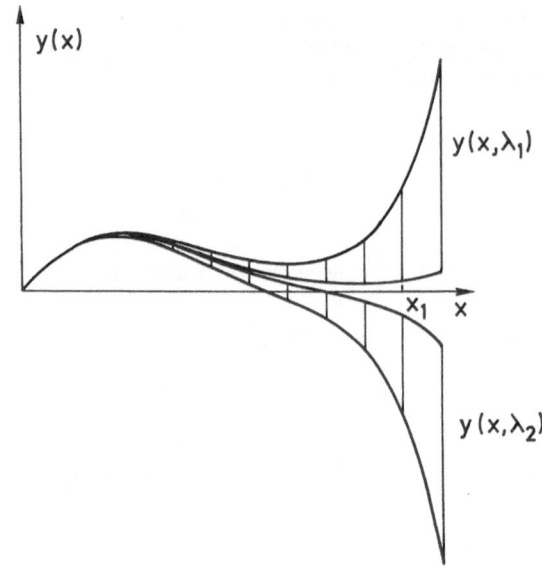

Figure 2: Typical behaviour of the inclusion of $y(x, \lambda)$ for values λ_1, λ_2 enclosing an eigenvalue. Even bad inclusions are sufficient to detect a sign change of $y(x)$ at x_1.

The problem should be well posed and it is assumed that the solution of the initial value problem exists for each $\lambda \in I$ and depends continuously on λ. This situation resembles the situation (7) together with (8).

Solving the initial value problem on a computer one may notice the same behaviour as observed above. Depending on λ the solution $y(x, \lambda)$ will tend to $-\infty$ or $+\infty$ and the changes of sign are associated with the eigenvalues of the problem. (see Fig. 2)

The following approach may be used for a stopping rule to decide the "sign" of $y(x)$ for large x. Assume that there exist constants x_0 and $M > 0$ (independent of $\lambda \in I$) such that

$$f(x) - \lambda > M \quad \text{for} \quad x > x_0 \tag{23}$$

If now

$$y(x_1, \lambda) > 0 \quad \text{and} \quad y'(x_1, \lambda) > 0 \quad \text{for} \quad x_1 > x_0 \tag{24}$$

then a simple comparison argument shows that $y(x, \lambda)$ will exhibit no further sign changes. The stopping rule condition (24) again can be checked on a computer using a verifying initial value problem solver (Lohner[3], Kaucher[2]).

The fact that sign changes in the behaviour of the solutions are associated with eigenvalues can again be proved under certain assumptions concerning the functions $f(x), g(\lambda), h(\lambda)$. The proof uses arguments similar to those in the case of the difference equation and will therefore not be given here.

3.3   Implementation

Only few examples have been checked to test the applicability of this method but the results obtained were quite good. Fig. 2 indicates that the sign decision again can be made even if the inclusion for the solution of the initial value problem is very large.

If a critical situation is encountered, meaning that zero is contained in the intervals for $y(x_1, \lambda)$ and $y'(x_1, \lambda)$ this again indicates that the problem has to be treated using higher accuracy. This leads to a simple automated adjustment of the stepsize and approximation order in the method of Lohner[2] or other methods.

Using a simple Picard iteration and interval polynomials with Tschebyscheff rounding to obtain solutions to the initial value problem the eigenvalues of the problem

$$y''(x) = (x - \lambda)y(x) \quad , \quad y(0) = 0 \quad y'(0) = 1 \tag{25}$$

could be computed up to 9 digits using 15 digit arithmetic. It should be mentioned that the value x_1 up to which the initial value problem has to be solved usually seems to be quite small and grows only slowly when the inclusion interval for the eigenvalue gets narrow.

All computations have been carried out by means of a microcomputer using a series 32000 processor together with a floating point unit. The programs were programmed in a language designed for easy use of interval arithmetic on this processor making use of the facilities of the floating point unit.

References

[1] Wimp, J.: Computation with Recurrence Relations ; Pitman Publishing, Applicable Mathematics Series XII, 1984.

[2] Lohner, R.J.: Enclosing the solutions of ordinary initial- and boundary- value problems ; in : Compterarithmetic Ed. by Kaucher,E.,Kulisch,U.,Ullrich,C., Teubner 1987.

[3] Kaucher, E., Miranker, W.L.: Self-Validating Numerics for Function Space Problems, Academic Press, New York, 1984.

[4] Perron, O.: Die Lehre von den Kettenbrüchen, Chelsea, 1950.

Dipl.-Ing. Martin Ohsmann
Lehrstuhl I für Mathematik
RWTH Aachen
Augustinerbach 2a
D-5100 Aachen
Federal Republic of Germany

II. Applications in the Technical Sciences

Computing, Suppl. 6, 91–98 (1988)

© by Springer-Verlag 1988

VIB — Verified Inclusions of Critical Bending Vibrations

A. Ams and W. Klein, Karlsruhe

Summary — Zusammenfassung

VIB — Verified Inclusions of Critical Bending Vibrations. In a joint project of the Institutes of Applied Mathematics and Technical Mechanics of the University of Karlsruhe a program system was developed for the calculation and animation of the bending vibrations of rotating shafts. The mechanical model of the rotor is based on the Bernoulli-Euler beam and evaluated by means of transfer matrices. By means of verified inclusions the eigenvalues can be determined with high accuracy up to high speeds of rotation.

VIB — Verifizierte Erschließungen kritischer Biegeschwingungen. In einem gemeinsamen Projekt der Institute für Angewandte Mathematik und für Technische Mechanik der Universität Karlsruhe wurde ein Programmsystem zur Untersuchung biegekritischer Drehzahlen rotierender Wellen entwickelt. Das zugrundeliegende mechanische Modell basiert auf der Bernoulli-Euler Theorie und wird mittels Übertragungsmatrizen ausgewertet. Über verifizierte Einschließungen werden Eigenfrequenzen mit einer hohen Genauigkeit auch für hohe Drehzahlbereiche berechnet.

1. Technical Backgrounds

A basic problem in technical dynamics is the investigation of bending vibrations of rotating shafts. They are excited by unbalanced masses and particularly dangerous for life time when the speed of rotation reaches the neighbourhood of the natural frequencies of the mechanical system. Therefore, the knowledge of such critical speeds is important for controlling the speeding up of the rotor. In case that speed fluctuations are occurring, an optimal layout of the rotor configuration and the bearing positions has to be found in order to get bending vibrations with smallest possible amplitudes in a wide frequency range.

In general, technical rotors are determined and influenced by many parameters coming up from the dimensioning of shaft lengths and cross sections as well as from the positioning of the rotor bearings and associated material characteristics. Since the calculation of natural frequencies is nonlinear, an optimal design of the entire system is a highly nonconvex problem, not solvable in general. Therefore, one has to try to get suboptimal solutions by looking for the influences of the most important parameters in a purely experimental way by numerical simulations. These calculations require a dialog-oriented program system with interactive input modes and graphical controlling. Furthermore, efficient algorithms and a tool for writing compact programs are important for time optimal handling of the required variation of all rotor data.

2. Program Survey

In a joint project of the Institute of Applied Mathematics and the Institute of Technical Mechanics a program system for the calculation of bending-critical rotation speeds by means of transfer matrices was developed. This VIB-program (Verified Inclusions for the critical **B**ending vibrations of a rotor beam) allows the user to construct the rotor configuration with different bearing types and to simulate the natural bending vibrations of the rotor shaft. The program is written in PASCAL-SC including verified calculations of the natural frequencies and associated modes and uses the computer arithmetic as defined in [4]. It runs on the personal computer KWS SAM 68 K with ca. 1 MB main storage and a graphic adaptor. The output of listings and graphics can be made on screen, printer or plotter. VIB is an extremely dialog-oriented program system. The user controls the calculations and program bifurcations by means of function keys on the keyboard of the computer. Similar as shown in figure 1, applications of window techniques simultaneously give graphical and numerical information on screen. Every possible branching is shown in a menu on the bottom of the screen.

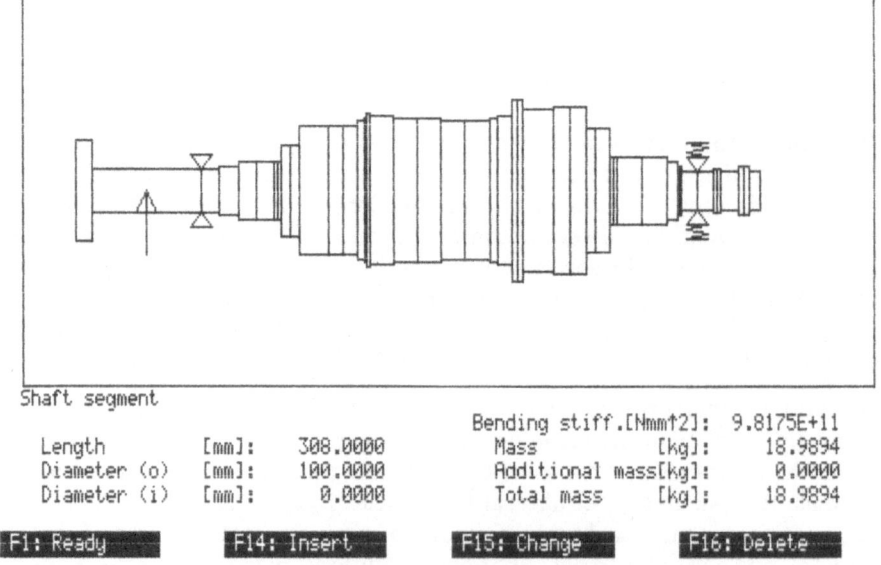

Shaft segment

			Bending stiff.[Nmm↑2]:	9.8175E+11
Length [mm]:	308.0000		Mass [kg]:	18.9894
Diameter (o) [mm]:	100.0000		Additional mass[kg]:	0.0000
Diameter (i) [mm]:	0.0000		Total mass [kg]:	18.9894

| F1: Ready | F14: Insert | F15: Change | F16: Delete |

Figure 1. Computer aided design of rotors and bearings

In the CAD-part of the VIB program the construction of the rotor is done either by means of file systems or interactively on the keyboard of the personal computer. Segments of the rotor shafts are indicated by a cursor. They can interactively be inserted or cancelled by function keys. The relevant parameters such as segment lengths, inner or outer circle diameters and positive or negative additional masses are organized in a rotor editor mode calculating the associated masses and bending stiffness values of the indicated segment. Subsequently, rotor bearings can be inserted at the position by function keys (see fig. 1).

The CAE-part of VIB performs a modal analysis of the rotor system by means of transfer matrices. The calculation of bending vibrations starts with an iterative determination of the natural frequencies for a chosen range of rotation speeds and frequency steps. The verified inclusions are shown on the bottom of the residual frequency determinant. Subsequently, the eigenforms of each calculated natural frequency are calculated and represented on the screen. Similar representations are available for the angles of beam inclination, for the distribution of the moments and the shear forces. A specifically adapted conversion into integer graphic units leads to a fast image sequence of the shaft deformations and inner force distributions so that one gets impressive animations of the time-dependent behaviour of the bending rotor vibrations.

3. Mechanical Model

The theory of Bernoulli-Euler beams leads to the homogeneous partial differential equation of the form

$$EI\, w_{xxxx}(x, t) + \mu w_{tt}(x, t) = 0 \qquad (0 \leq x \leq 1) \tag{1}$$

which describes the free bending vibrations $w(x, t)$ of a cylindric shaft segment in dependence of space and time. Accordingly, x is the local space coordinate of the segment, EI is its bending stiffness and μ is the mass per unit length l in one segment. The subscripts denote derivations with respect to the space coordinate x and the time variable t. The above differential equation is separated by means of the isochronic set-up $w(x, t) = W(x) \cos \omega t$. Subsequently, it is integrated in the space domain and fitted to the boundary conditions at the beginning of each rotor segment $i \in I := \{1, 2, \ldots, n\}$.

These four boundary conditions are noted in a column matrix z_i and given by the initial states of the beam segments

$$
\begin{aligned}
\text{deflection amplitude} \quad & W_i = W(0), \\
\text{angle of inclination} \quad & \psi_i = -W_x(0), \\
\text{bending moment} \quad & M_i = -EIW_{xx}(0), \\
\text{beam shear force} \quad & Q_i = -EIW_{xxx}(0).
\end{aligned}
\tag{2}
$$

The next state z_{i+1} at $x=l$ is calculated by means of the transfer matrix U_i.

$$
\begin{bmatrix} -W_{i+1} \\ \psi_{i+1} \\ M_{i+1} \\ Q_{i+1} \end{bmatrix} = \begin{bmatrix} C_p & S_p/\lambda & C_n/\lambda^2\,EI & S_n/\lambda^3\,EI \\ \lambda S_n & C_p & S_p/\lambda EI & C_n/\lambda^2\,EI \\ \lambda^2\,EIC_n & \lambda EIS_n & C_p & S_p\lambda \\ \lambda^3\,EIS_p & \lambda^2\,EIC_n & \lambda S_n & C_p \end{bmatrix} \begin{bmatrix} -W_i \\ \psi_i \\ M_i \\ Q_i \end{bmatrix}
$$

$$
z_{i+1} \qquad = \qquad\qquad\qquad U_i \qquad\qquad\qquad\qquad z_i,
$$

Herein, S and C are abbreviations of the so-called Rayleigh functions

$$
C_{p(n)} = \tfrac{1}{2}(\cosh \lambda 1 \underset{(-)}{+} \cos \lambda 1),
$$

$$
S_{p(n)} = \tfrac{1}{2}(\sinh \lambda l \underset{(-)}{+} \sin \lambda l), \qquad \lambda^4 = \frac{\mu\omega^2}{EI}, \tag{4}
$$

which follow form the space integration of the separated equation of motion. The parameter λ is the eigenvalue of the separated problem containing the critical rotation speed ω of interest.

Applying the matrices U_i for $i\in I$, the transfer calculation starts with z_0 at the left side of the rotor and ends with z_n on the right side. This results in a recurrence algorithm giving the following global 4×4 transfer matrix of the rotor:

$$
z_n = (\prod_{i\in I} U_i)\, z_0, \qquad \det U^*(\lambda) = 0. \tag{5}
$$

Two elements of z_0 and z_n are unknown. The other are always given and vanishing according to the global boundary conditions of each rotor end. Consequently, the eigenvalue problem is reduced to a 2×2 matrix U^*. Its determinant has to be zero for nontrivial solutions of the two initial state elements, mentioned above. Further details on extensions to inner boundary conditions or higher mechanical models are given in [1], [2] and [3].

4. Numerical Difficulties

The determination of the function values (5) at discrete points in a given interval makes an iterative computation of the eigenvalues and appropriate eigenvectors of the rotor system possible. Extending the considered interval to higher frequencies leads to strange numerical behaviour of the bending vibration in contradiction to the expected physical or mechanical performance.

Because of the fact that the applied mechanical and mathematical models are suitable for these frequencies, the source of error was supposed in the computer arithmetic used. These high frequency results gave rise to a certain distrust in the determination of the eigenvalues for lower frequencies also.

During the cooperation between the Institute of Technical Mechanics and the Institute for Applied Mathematics the algorithm for the determination of the

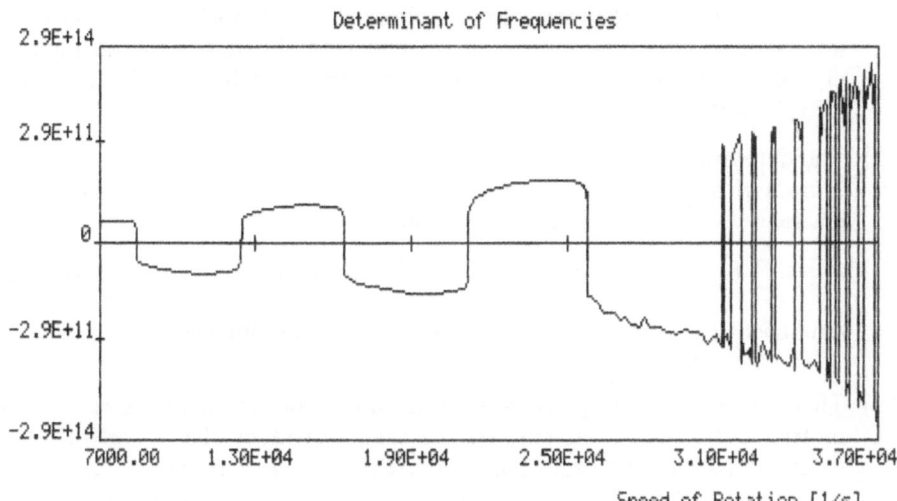

Figure 2. Residual frequency determinant

eigenvalues was carefully analyzed. The method of transfer matrices essentially consists of three steps. Given a stepsize and an interval, these three steps have to be carried out for each discrete frequency $f_i = f_0 + i * h$, $i \in I$. As shown above, each rotor segment S_i is represented by a transfer matrix U_i, $i \in I$.

1) First the coefficients of each transfer matrix U_i (3) have to be determined. The solution of the partial differential equation (1) causes the dependency of the coefficients on trigonometric and hyperbolic functions, more precisely on the so-called Rayleigh functions (4).

2) Then the boundary state of each rotor segment has to be determined by the product of its representing transfer matrix U_i and the boundary state of the preceding segment $z_{i+1} = U_i * z_i$, $i \in I$ (see (3) and (5)).

3) In the last step the determinant of the resulting product matrix has to be computed and has to be compared with the determinant of the preceding frequency f_{i-1}. In the case of opposite signs, the interval $[f_{i-1}, f_i]$ contains an eigenvalue of the rotor which is determined iteratively more exactly.

The implementation of this algorithm on a computer as realized in a program system called TRAMA [1] leads to the numerical problems shown in fig. 2. They arise during all three steps.

ad 1) When computing the Rayleigh functions (4) using a floating point arithmetic with 13 decimal digits and using arguments $x > 30$, identical results are obtained for all of the following expressions:

$$(\sinh(x) + \sin(x))/2 = (\cosh(x) + \cos(x))/2$$
$$= \sinh(x)/2 \qquad = \cosh(x)/2$$
$$= \exp(x)/4$$

That means, the trigonometric functions as well as the exponential function with negative exponent will be neglected. Therefore the coefficients of the transfer matrices cannot be determined with sufficient accuracy for this problem.

ad 2) The determination of an exact product of matrices is a very old problem in numerical computation. Computing the product $A := B * C$ of only two matrices B and C leads to $2n-1$ rounding operations for each element

$a_{i,j} = \sum\limits_{k=1}^{n} b_{i,k} * c_{k,j}$, $i, j \in I$. Of course, in our special case the product of

n matrices $\prod\limits_{i \in I} U_i$ as described in (5) increases rounding errors enormously.

ad 3) There are different ways of computing the determinant of a given $(n \times n)$ matrix, but all methods employ the scalar product of two vectors. For example, the LU-factorization of a matrix followed by the computation of the product of the main-diagonal elements of the matrix L leads to the formula

$$l_{i,j} = \left(a_{i,j} - \sum\limits_{k=1}^{i-1} l_{i,k} * u_{k,j} \right) \bigg/ l_{i,i}, \quad i, j \in I$$

for the elements of the lower-triangular matrix L. Again similar to 2), $2i-3$ rounding operations are needed for the determination of each matrix element of L.

5. Numerical Inclusions

Each of the three described steps of the modal analysis (see above) was improved by using the computer arithmetic as defined in [4] in order to get verified inclusions for the eigenvalues of the rotor.

ad 1) Two dissertations by Braune and Krämer [5] on the computation of standard functions with arbitrary precision and verified results were used. The Rayleigh functions (4) and therefore all coefficients of the transfer matrices U_i, $i \in I$ (3), are determined up to an accuracy of 52 decimal digits. The last 13 digits are enclosed in an interval which contains the true mathematical result.

ad 2) There is a simple way to avoid repeated matrix multiplications. By using only one band-shaped matrix, the transfer matrices U_i can be stored as block matrices. The resulting linear system of equations is solved with the residuum technique and with the exact scalar product of two vectors. This exact scalar product function, which is part of PASCAL-SC and described in [4] and [6], involves only one rounding of the result to the preceding, succeeding or nearest floating-point number, in contrast to the $2n-1$ rounding operations of the normal scalar product.

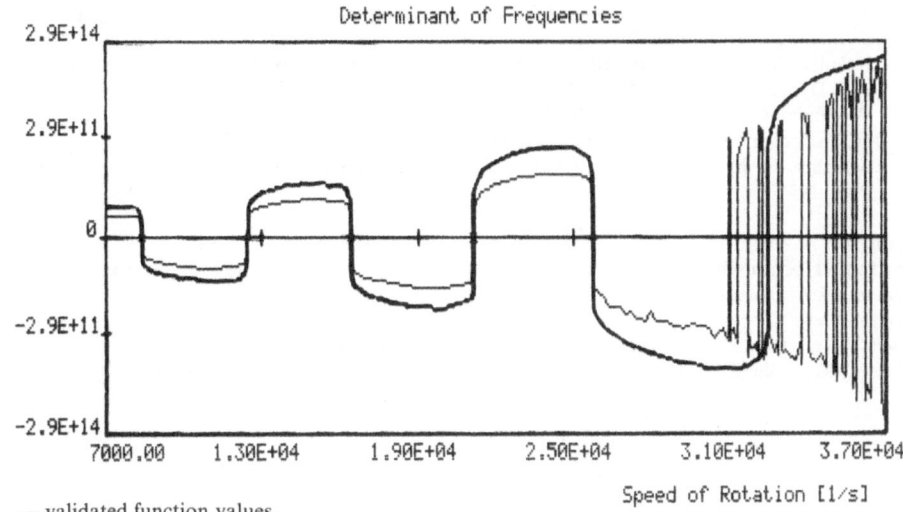

—— validated function values.

Figure 3.
Comparison between classical numerics and verified results

ad 3) PASCAL-SC enables working with external files. For our program system VIB, an existing program for the determination of the *LU*-factorization of a given (*n* × *n*) matrix with predetermined precision and verified results was used. Computing the product of the main-diagonal elements of *L*, the determinant of the matrix is determined with verified result up to the chosen accuracy.

These improvements in the new program VIB produced, in contrast to the old program TRAMA [1], the results of the next picture.

With the new program, the eigenvalues for high frequencies as well as for lower frequencies can be determined with high accuracy. Furthermore, the results are verified to be correct.

Acknowledgment

Special thanks for their support and helpful ideas go to Prof. E. Kaucher and Prof. W. Wedig.

References

[1] Zürn, O.: Berechnung biegekritischer Drehzahlen von Rotoren mit Hilfe von Übertragungsmatrizen an einem IBM PC, Project 1986, Institute of Technical Mechanics, University of Karlsruhe.
[2] Wedig, W., Ams, A.: Demonstration Programs in Technical Mechanics, Computer Applications University Karlsruhe 3, March 1987.
[3] Pestel, E., Leckie, F.A.: Matrix Methods in Elastomechanics, Mc Graw-Hill, New York 1963.
[4] Kulisch, U., Miranker, W.: Computer Arithmetic in Theory and Practice, Academic Press, New York, 1981.

[5] Braune, K.: Standard Functions for Real and Complex Point and Interval Arguments with Dynamic Accuracy, this Volume.
Krämer, W.: Inverse Standard Functions for Real and Complex Point and Interval Arguments with Dynamic Accuracy, this Volume.

[6] Bohlender, G., Rall, L.B., Ullrich, Ch., Gudenberg, J.: PASCAL-SC: Wirkungsvoll programmieren, kontrolliert rechnen. Bibliographisches Institut, Mannheim/Wien/Zürich, 1986.

[7] Kaucher, E., Miranker, W.: Self-Validating Numerics for Function Space Problems, Academic Press 1984

Dipl.-Ing. A. Ams   Dipl.-Math. W. Klein
Institut für Technische Mechanik   Institut für Angewandte Mathematik
Universität Karlsruhe   Universität Karlsruhe
Kaiserstrasse 12   Kaiserstrasse 12
D-7500 Karlsruhe 1   D-7500 Karlsruhe 1
Federal Republic of Germany   Federal Republic of Germany

Computing, Suppl. 6, 99—110 (1988)

Computing
© by Springer-Verlag 1988

Stability Test for Periodic Differential Equations on Digital Computers with Applications

D. Cordes, Karlsruhe

Abstract – Zusammenfassung

Stability Test for Periodic Differential Equations on Digital Computers with Applications. For linear and nonlinear ordinary initial value problems with periodic coefficient functions, a sufficient enclosure test is derived. The successful execution on digital computers guarantees the stability of all solutions in the linear case or the boundedness of all perturbed solutions starting in an admitted finite neighborhood of the initial data in the nonlinear case. Presented applications are concerned with vibrations of gear drives. An implementation of the test using FORTRAN and ACRITH is discussed.

Nachweis der Stabilität bei periodischen Differentialgleichungen auf dem Rechner mit Anwendungen. Für lineare und nichtlineare gewöhnliche Anfangswertprobleme mit periodischen Koeffizientenfunktionen wird ein hinreichender Einschließungstest hergeleitet. Die erfolgreiche Durchführung des Tests auf digitalen Rechenanlagen garantiert die Stabilität aller Lösungen im linearen Fall und die Beschränktheit aller gestörten Lösungen im nichtlinearen Fall, die in einer endlichen, zulässigen Umgebung des Anfangswertes beginnen. Die vorgestellten Anwendungen befassen sich mit Getriebeschwingungen. Eine Implementierung des Tests unter Verwendung von FORTRAN und ACRITH wird behandelt.

1 Introduction

The following type of initial value problems will be considered:

$$y'(t) = f(t, y(t)) \text{ for } t > t_0,$$

$$y(t_0) = y_0, \ y(t) \in \mathbf{R}^n \text{ for } t \geq t_0,$$

(1)

with $f(t + T, .) = f(t, .)$ for all $t \geq t_0$ where $T \in \mathbf{R}^+$ denotes the common period of the coefficient functions.

A linear system of ordinary differential equations of order n with periodic coefficient functions may possess nonempty domains of stability in a parameter space such that every solution is bounded irrespective of the choice of the initial vector $y_0 \in \mathbf{R}^n$. Domains of stability may be empty as, e.g., in the case of the "totally unstable" system [5, p.76]

$$u''(t) + a^2 \cdot u(t) + e \cdot \sin(\Omega \cdot t) \cdot v(t) = 0,$$
$$v''(t) + b^2 \cdot v(t) + e \cdot \cos(\Omega \cdot t) \cdot u(t) = 0,$$

(2)

with a,b,e,$\Omega \in \mathbf{R}$.

In general, the exact determination of intervals of stability (or instability) is difficult. Therefore, computational tests for the verification of stability on digital computers are of particular interest.

In applications, domains of instability are related to parametric resonance (if $n \geq 2$) or combinational resonance (if $n > 2$). According to the Floquet theory [5, p.55-59] a point in the parameter space is stable if the spectral radius $\varrho(Y(T))$ of the fundamental matrix Y(t) at time $t = T$ is less than one. In stable neighborhoods of the boundary of the domains of stability which may be of considerable size, the spectral radius of $Y(T)$ is close to one. Consequently, reliable practical applications of the criterion $\varrho(Y(T)) < 1$ are extremely difficult due to the approximate character of available numerical methods for the computation of $Y(T)$ and the determination of the spectral radius of $Y(T)$. This is enhanced by the sensitivity of the eigenvalues with respect to (small) changes in the elements of $Y(T)$.

In the nonlinear case practical methods for the determination of the boundedness of solutions rest on uncontrolled approximations such as the linearization of the problem or the truncation of an iteration. Consequently, there is an urgent need for reliable computational tests for the determination of the boundedness of solutions.

Historically, ordinary differential equations with periodic coefficients were first investigated by

- Mathieu (1868) [13] in his construction of a solution of the wave equation via separation of variables in elliptical coordinates which yielded the classical Mathieu equation

$$u''(t) + (a + b \cdot \cos(2 \cdot t)) \cdot u(t) = 0 \tag{3}$$

  with $a, b \in \mathbf{R}$, and by

- Hill (1877) [8] in his investigation of the perigee of the moon which yielded a class of Hill equations

$$u''(t) + \left(\Phi_0 + \sum_{n=1}^{\infty}(\Phi_n \cdot \cos(2 \cdot n \cdot t))\right) \cdot u(t) = 0 \tag{4}$$

  with $\Phi_i \in \mathbf{R}$, $i = 0, 1, \ldots$.

The complexity of the problems under discussion is exhibited by the fact [15, p.19] that the addition of a linear term with a periodic coefficient function to a stable autonomous system of ordinary differential equations may destabilize the extended system, and the addition of such a term to an unstable autonomous system may stabilize the extended system. Examples for these destabilizing and stabilizing effects are the suspended and the erected pendulum both with periodic oscillating suspension, respectively.

Ordinary differential equations with periodic coefficient functions frequently occur in mathematical simulation of vibrations in gear drives and vibrations due to periodic changes of parameters, e.g., in robotics as well as in wide classes of problems in mechanical engineering.

2   Stability Test for Linear Problems

The following first order system of linear ordinary differential equations will be investigated:

$$y'(t) = (A + B(t)) \cdot y(t) \text{ for } t \in \mathbf{R}^+, \tag{5}$$

with

$$A \in L(\mathbf{R}^n), \ B : \mathbf{R} \to \mathbf{R}^n, \ B \text{ is continuous,}$$
$$B(t + T) = B(t) \text{ for all } t \geq 0, \ T \in \mathbf{R}^+ \text{ is fixed}$$

and initial data

$$y(0) = y_0 \in S_0 := \delta \cdot [-1, 1] \cdot d,$$
$$\delta \in \mathbf{R}^+, d = (1, \dots, 1)^T \in \mathbf{R}^n. \tag{6}$$

Due to the affine mapping of S_0 by means of the fundamental matrix $Y(t)$ the set of solutions

$$S_t := \{y(t)|y'(t) = (A + B(t)) \cdot y(t), \ y_0 \in S_0\} \subset \mathbf{R}^n \tag{7}$$

for any $t > 0$ is a parallelepiped containing the zero vector. By use of [12], which is the essential basis for the current implementation, the components of the column vectors $c^j := (c_{1j}, c_{2j}, \dots, c_{nj})^T \in \mathbf{R}^n$, $j = 1, \dots, n$, of matrix $C := (c_{ij}) := Y(T)$ are enclosed with high precision. The successful execution of the computations yields an interval matrix $[C]$ with elements $[c_{ij}] := [\underline{c}_{ij}, \overline{c}_{ij}]$, $i, j = 1, \dots, n$, which are verified enclosures of the elements c_{ij} of C. For the exact solution $y(t)$ of (5), (6) there holds

$$y(T) = Y(T) \cdot y(0) = C \cdot y_0 \in [C] \cdot S_0 = \delta \cdot ([-1, 1] \cdot [C]) \cdot d$$
$$= \delta \cdot [-1, 1] \cdot (\textstyle\sum_{j=1}^n | [c_{1j}] |, \dots, \sum_{j=1}^n | [c_{nj}] |)^T, \tag{8}$$

where $| [c_{ij}] | := \max\{|\underline{c}_{ij}|, |\overline{c}_{ij}|\}$.

Consequently, the row sum criterion

$$\| \,|[C]|\, \|_\infty = \max_{1 \leq i \leq n} \sum_{j=1}^n | [c_{ij}] | < 1 \tag{9}$$

implies the validity of $S_T \subset S_0$. Because of the oscillatory solutions of (5), the parallelepiped S_T is in a rotated position as compared with the interval S_0. Consequently, the inequality (9) generally is not satisfied. However, there may exist a number $k \in \mathbf{N}$ such that for $t^* := k \cdot T$ there holds $S_{t^*} \subset S_0$. Since $y(k \cdot T) = C^k \cdot y_0$ for all $y_0 \in S_0$, the pertinent "enclosure condition"

$$\| \,|[[C]^k]|\, \|_\infty = \max_{1 \leq i \leq n} \sum_{j=1}^n | [c_{ij}^{(k)}] | < 1 \tag{10}$$

where $C^k \in [[C]^k] := ([c_{ij}^{(k)}])$ for all $C \in [C]$, implies that

$$y(t^*) \in S_{t^*} \subset \delta \cdot [-1, 1] \cdot d = S_0 \text{ for all } y_0 \in S_0, \tag{11}$$

analogously to (9). A continuation of this construction yields $y(m \cdot t^\ast) \in S_0$ for all $m \in \mathbf{N}$ and all $y_0 \in S_0$. This implies the boundedness of the set of solutions S_t for all $t > 0$, see [6]. The constructive verification of the condition (10) is a practical implementation of the stability condition $\varrho(Y(T)) < 1$ which is due to the Floquet theory [7]:

- a fundamental matrix $Y : \mathbf{R} \to \mathbf{R}^n$ of (5) satisfies

$$Y(t + T) = Y(t) \cdot C \tag{12}$$

  for all $t \geq 0$ and $C \in L(\mathbf{R}^n)$. If $Y(0)$ is the identity matrix $I \in L(\mathbf{R}^n)$, then $C = Y(T)$.

- if and only if $\varrho(C) < 1$, then there holds asymptotic stability, i.e., all solutions $y(t)$ of (5) approch 0 as $t \to +\infty$.

Here and in the following, brackets '[ ]' identify an interval-type result of the evaluation of a mathematical expression. Thus, the notation $[[C]^k]$ in (10) stands for a safe enclosure of $[C]^k$ where any local error is taken into due account by means of an enlargement of the component intervals if necessary.

The basic idea of the following version of an algorithm for the determination of a sequence of interval matrices $[[C]^i], i = 1, 2, \ldots$, goes back to Krückeberg and Moore. This idea, which is also essential in the enclosure method developed in [12], consists of the premultiplication of an interval matrix, say $[A]$, by the inverse of a suitably chosen point-matrix $P \in [A]$.

Algorithm   (13)

initialize matrices	$P_0 := I,\ [B_0] := I,\ [[C]^1] := [C],\ [R_0] := [C]$	
choose a point-matrix	$P_i \in [R_{i-1}]$	
enclose the inverse	$[Q_i] := [P_i^{-1}]$	
multiply matrices	$[B_i] := [[[Q_i] \cdot [R_{i-1}]] \cdot [B_{i-1}]]$	$i = 1, 2, \ldots$
multiply matrices	$[R_i] := [[C] \cdot P_i]$	
determine enclosure	$[[C]^{i+1}] := [[R_i] \cdot [B_i]]$	

The execution of this algorithm requires the determination of an interval enclosure $[P_i^{-1}]$ for the inverse of the point-matrix $P_i \in L(\mathbf{R}^n)$ in every step of the algorithm. The determination of such an enclosure usually is very time consuming. In order to improve efficiency a simplified algorithm may be employed which chooses a new point-matrix P_i in selected steps only and, otherwise, uses straightforward interval arithmetic evaluations of matrix products.

The interval matrices $[[Q_i] \cdot [R_{i-1}]]$ are "almost diagonal" for all $i \in \mathbf{N}$. By construction the interval matrices $[B_i]$, $i \in \mathbf{N}$, are "almost diagonal", too, and do not suffer from over-estimations due to rotations which are induced by $Y(T) \in [C]$. The interval matrices $[R_i]$ gather the rotational effects of $[C]$ and form the essential part in the determination of $[[C]^{i+1}]$. The interval matrices $[[C]^j]$, $j \in \mathbf{N}$, are not used in the determination of $[[C]^k]$ for $k > j$.

Computational experience for the determination of bounds for $\varrho(Y(T))$ showed that in many cases an estimation of the norm of the interval matrix $[\,[C]^k]$ which has been determined by use of "naïve" interval multiplications, may already satisfy condition (10). The following algorithm may be used for a "fast" determination of large powers of $[C]$.

Algorithm $\hspace{9cm}$ (14)

$$
\boxed{
\begin{aligned}
&\text{initialize} \quad [C]^{(1)} := [C] \\
&\text{determine} \quad [C]^{(j+1)} := [C]^{(j)} \cdot [C]^{(j)} \,,\; j = 1, 2, \ldots
\end{aligned}
}
$$

For any $j \in \mathbf{N}$ the interval matrix $[C]^{(j+1)}$ is an enclosure of the set of matrices $\{C^{2^j} : C \in [C]\}$. If the enclosure condition $\|\,|\,[C]^{(j+1)}|\,\|_\infty < 1$ is satisfied for a number $j \in \mathbf{N}$, then $k := 2^j$ in (10) and $t^* := 2^j \cdot T$ in (11). Of course, the value of $j \in \mathbf{N}$ may be large, especially, if $\varrho(Y(T))$ is close to one. The sequence of norms of matrices $[C]^{(j+1)}$ determined by the "fast" algorithm (14) may even diverge whereas the use of $[\,[C]^{i+1}]$ determined by algorithm (13) still succeeds in satisfying the enclosure condition (10).

In both algorithms (13) and (14), the symbol '·' stands for the usual interval arithmetic multiplication of real interval matrices (see [1] for details about interval arithmetic).

3   Stability Test for Nonlinear Problems

For a nonlinear initial value problem (1) with periodic coefficients of a common period $T \in \mathbf{R}^+$ and the admission of an initial interval $[y_0]$ of finite width, it is assumed that a component-wise enclosure of the set of solutions has been determined by use of [12] for $t \in [t_0, t_0 + k \cdot T]$ where $k \in \mathbf{N}$ is sufficiently large. It is further assumed that there is a $K \in \{1, 2, \ldots, k\}$ such that an enclosure condition is satisfied at $t^* := t_0 + K \cdot T$, i.e., the computed enclosure $[y(t^*)]$ is contained in the admitted initial interval $[y_0]$. Due to the enclosing character of $[y(t)]$ for all $t \geq t_0$ and all $y_0 \in [y_0]$, the set of solutions S_t is contained in the computed interval enclosure $[y(t)]$ for all $t \in I_0 := [t_0, t^*]$, and this is also true for any interval $I_m := [t_0 + m \cdot K \cdot T, t_0 + (m+1) \cdot K \cdot T]$ and all $m \in \mathbf{N}$. This establishes the boundedness of every solution of the initial value problem possessing admitted initial data $y_0 \in [y_0]$. In the linear case, the enclosure $[y(t)]$ is required only for $t \in [t_0, t_0 + T]$, i.e., for a time interval of the length of one period T, due to the Floquet theory.

For a nonlinear system of differential equations (1) with $f(t, 0) = 0$ for all $t > t_0$, the practical stability [11, p.109] of the trivial solution $y(t) = 0$ may be verified via the determination of a number $K \in \mathbf{N}$ such that $[y(t_0 + K \cdot T)] \subset [y_0]$ under the admission of all initial vectors $y_0 \in [y_0]$. If the lower and upper bounds of the interval elements $[y_{0,i}] := [\underline{y}_{0,i}, \overline{y}_{0,i}]$ of $[y_0]$ satisfy $\underline{y}_{0,i} < 0 < \overline{y}_{0,i}$ for $i = 1, \ldots, n$, then the existence of K ensures the boundedness of all admitted perturbed solutions for all $t \geq t_0$.

4 Implementation

Both the enclosure method described in [12] and the algorithms (13) and (14) have been implemented in FORTRAN on an IBM 4361. The arithmetic has been supported by the FORTRAN subroutine library ACRITH [9], which includes (among others) basic arithmetic subroutines for an optimal computer arithmetic according to [10], standard functions for interval and point arguments, and problem solving subroutines like function evaluation, and linear and nonlinear system solvers.

The required computer arithmetic for the implementation of verifying algorithms must provide the controlled rounding of results of arithmetic operations and the determination of an exact dot product. Up to now, only the programming languages PASCAL-SC [4] and FORTRAN-SC [3] both developed in the Institute for Applied Mathematics at the University of Karlsruhe support these arithmetic requirements by language concepts. Unfortunately, these comfortable programming tools are not yet widely distributed, thus restricting the portability of developed source programs. On the other hand, the FORTRAN subroutine libraries ACRITH [9] and ARITHMOS [2] provide an easily accessible interface to the required features of an optimal computer arithmetic in the sense of [10].

The implementation of (13) and (14) is straightforward if the subroutine libraries ACRITH and ARITHMOS are used, since they provide individually reusable subroutines which handle the multiplication of interval matrices and the determination of a verified enclosure of the inverse of a point matrix.

The more cumbersome part is the implementation of the method for the determination of a verified enclosure of the solution of an initial value problem as described in [12]. This is enhanced by the desire to create a more general and interactive program which handles not only a special class of problems. The following list of aspects was taken into account:

- almost mathematical notation for the input of a first or second order system of explicit ordinary differential equations,

- interactive input of initial data,

- interactive input of method parameters like step size and order,

- allowing right-hand side functions f in the system of differential equations which involve the following operators and functions
    - unary +, unary −, +, −, *, /
    - ** = exponentiation with constant exponent
    - sin, cos, sinh, cosh, log, exp
    - sqr = square, sqrt = square root

  and the definition of piecewise constant composite functions with a finite number of points of discontinuity, and

- input of real and interval data in a decimal and a machine dependent (hexadecimal) notation.

Piecewise constant composite functions are used as a tool for the construction of arbitrary composite functions. The following examples may illustrate the use of piecewise constant composite functions. Suppose $s(t)$ is defined by

$$s(t) = \left\{ \begin{array}{ll} +1 & 0 \leq t < 1 \\ -1 & 1 \leq t < 2 \end{array} \right.$$

then the subsequently defined functions $f(t)$ and $g(t)$

$$f(t) := |\sin(\pi \cdot t)| \text{ for } 0 \leq t < 2,$$

$$g(t) := \left\{ \begin{array}{ll} 2 \cdot t^2 & 0 \leq t < 1 \\ 2 \cdot \sqrt[3]{t} & 1 \leq t < 2 \end{array} \right.$$

may be expressed equivalently by

$$f(t) = s(t) * \sin(\pi \cdot t),$$
$$g(t) = (s(t) + 1) \cdot t^2 + (1 - s(t)) \cdot \sqrt[3]{t}.$$

The admittance of piecewise constant composite functions requires that the points of discontinuity $\tilde{t}$ are points of the chosen time grid and possess an exact representation in the floating-point number system of the computer. These special points must be taken into account when step sizes for the one-step method in [12] are determined.

In the current implementation the step size $h_{j+1} := t_{j+1} - t_j$ is chosen less than or equal to a (maximum) step size h_{max} which is prescribed by the user, and such that all points $\tilde{t}$ induced by the composite functions are grid points. The subsequently described algorithm has been implemented for the determination of the step sizes $h_{j+1}, j + 1 \in \mathbf{N}$.

Algorithm (15)

(15a) Determine next required grid point $\tilde{t} > t_j$.
(15b) If $t_j + h_{max} \geq \tilde{t}$, then $h_{j+1} := \tilde{t} - t_j$.
(15c) If $t_j + h_{max} < \tilde{t}$ and $t_j + 1.5 \cdot h_{max} \geq \tilde{t}$, then $h_{j+1} := h_{max}/2$.
(15d) Otherwise, $h_{j+1} := h_{max}$ is used.

If no further point $\tilde{t}$ is found in (15a), then $\tilde{t}$ is set to the final time t_{end} which is prescribed by the user at program initialization. Step (15b) forces the point $\tilde{t}$ to be a grid point of the one-step method. In (15c) and (15d), the step size h_{j+1} may be chosen such that the quotients $\frac{h_{j+1}}{k}$, $k = 1, 2, \ldots, p$, are representable numbers in the underlying floating-point number system of the digital computer. Such a choice of h_{j+1} reduces the number of rounding operations required for the evaluation of the terms of the Taylor expansion. For the hexadecimal representation $0.m \cdot 16^e$ of the step size h_{j+1} determined by the FORTRAN implementataion of (15c) and (15d), the fractional part m is chosen to be an integer multiple of the (hexadecimally represented) number M:=0.63BAF3ED000000. Then, the value of h_{j+1} is at least divisible by the integer numbers $1, 2, \ldots, 26$ without rounding error provided the determination of the integer multiple of M did not cause a round-off error.

Due to the different bases of floating-point number systems used for the data input (decimal) and the computations on the digital computer (hexadecimal), conversion routines for the input and the output of data are required. In practical applications this will turn a problem with non-interval data into a problem involving interval data. For the used method [12] this is no restriction, since interval data is allowed for any parameter in the problem as well as for the prescribed inital data. The program output of the implemented method consists of the verified enclosure $[y(t)]$ of the solution $y(t)$ of the ordinary initial value problem at grid points t_j, and, optionally, a continuous representation of the enclosure for every time interval $[t_j, t_{j+1}]$ via a polynomial with interval coefficients

$$y(t) \in [y_j] + (t - t_j) \cdot [y_j'] + \ldots + \frac{(t - t_j)^{p-1}}{(p-1)!} \cdot [y_j^{(p-1)}] + \left(\frac{t - t_j}{h_{j+1}}\right)^p \cdot [z_{j+1}].$$

Here, the interval $[z_{j+1}]$ denotes the enclosure of the remainder term for all $t \in [t_j, t_{j+1}]$.

Within the FORTRAN program all formulas are internally represented by an integer indexed array structure which allows a straightforward numerical evaluation of the function $f(y)$ and its derivatives. The required partial derivatives $\frac{\partial f(y)}{\partial y}$ are formally determined once in the beginning of the execution of the program.

The mathematical standard functions used within the program accept point and interval arguments yielding verified enclosures of high accuracy. Of course, in connection with this program a representation of the required standard functions by first order systems of differential equations is possible, such that special subroutine entries seem to be unnecessary. However, such a replacement finally leads to a significantly smaller step size in the algorithm, which, in turn, results in a loss of precision in the determined enclosure $[y(t)]$ compared with the results obtained by direct calls to the standard functions of the ACRITH subroutine library.

Most important and most frequently used are those subroutines which allow manipulation of the long accumulators. In particular, the (cumulative) addition of dot products to an accumulator significantly simplifies programming. The long accumulators always hold the exact intermediate results without any round-off errors. Conversion of the accumulator contents to a (machine dependent) floating-point number representation is done only once, eventually combined with a controlled rounding operation. Thus, instead of up to (2n-1) round-off errors for dot products of real vectors with n elements, there is only one error due to controlled rounding if a long accumulator is used. Moreover, any intermediate exponent overflow or exponent underflow condition and the cancellation of leading digits of the mantissa do not cause a loss of accuracy in the final result.

5   Applications

The following problem comes from vibrational analysis of gear drives. The mathematical model is given by a first order system of explicit linear ordinary differential equations with periodic coefficients

$$y'(t) = (A + B(t)) \cdot y(t) \text{ for } t \in \mathbf{R}^+ \tag{16}$$

with $B(t + T) = B(t)$ for all $t \geq 0$ and fixed $T \in \mathbf{R}^+$. Required are verified statements about the stability of the trivial solution $y(t) = 0$. The periodicity of $B(t)$ is due to the periodically changing number of teeth in contact in any pair of mated gears.

Vibrations of three spur-type gears are simulated by means of a system (16) of order $n = 6$. The original model of three differential equations of second order which correspond to three rotational degrees of freedom

$$
\begin{aligned}
J_1 \cdot y_1'' + (d_{12} \cdot r_1^2 + d_1) \cdot y_1' + d_{12} \cdot r_1 \cdot r_2 \cdot y_2' \\
+ (k_{12} \cdot r_1^2 + k_1) \cdot y_1 + k_{12} \cdot r_1 \cdot r_2 \cdot y_2 = M_1, \\
J_2 \cdot y_2'' + d_{12} \cdot r_1 \cdot r_2 \cdot y_1' + (d_{12} + d_{23}) \cdot r_2^2 \cdot y_2' - d_{23} \cdot r_2 \cdot r_3 \cdot y_3' \\
+ k_{12} \cdot r_1 \cdot r_2 \cdot y_1 + (k_{12} + k_{23}) \cdot r_2^2 \cdot y_2 - k_{23} \cdot r_2 \cdot r_3 \cdot y_3 = 0, \\
J_3 \cdot y_3'' - d_{23} \cdot r_2 \cdot r_3 \cdot y_2' + (d_{23} \cdot r_3^2 + d_3) \cdot y_3' \\
- k_{23} \cdot r_2 \cdot r_3 \cdot y_2 + (k_{23} \cdot r_3^2 + k_3) \cdot y_3 = M_3,
\end{aligned}
\tag{17}
$$

has been transformed into a first order system by introducing additional differential equations for

$$
\begin{aligned}
y_4(t) &:= y_1'(t), \\
y_5(t) &:= y_2'(t), \\
y_6(t) &:= y_3'(t).
\end{aligned}
$$

In (17), the appearing terms stand for moments of inertia J_i, radii r_i, generalized coordinates y_i, $i = 1, 2, 3$, damping $d_{..}$, stiffness $k_{..}$, and external loads M_1, M_3. With the exception of k_{12} and k_{23} all terms are assumed to have constant values. The time-dependent functions $k_{12}(t)$ and $k_{23}(t)$ represent the stiffness of the teeth of a pair of mated gears. The pertinent stiffness functions oscillate periodically according to the number of teeth in contact. Between these functions a constant phase shift φ is assumed such that $k_{12}(t)$ and $k_{23}(t)$ may be expressed by

$$
\begin{aligned}
k_{12}(t) &:= k_{120} + k_{121} \cdot \sin(2\pi \cdot \omega \cdot t) \\
k_{23}(t + \varphi) &:= k_{230} + k_{231} \cdot \sin(2\pi \cdot \omega \cdot t + \varphi)
\end{aligned}
\tag{18}
$$

with $0 \leq \varphi < T := \frac{1}{\omega}$ and fixed values for $k_{120}, k_{121}, k_{230}, k_{231}, \omega \in \mathbf{R}^+$. The number ω stands for the frequency of teeth contacts of the mated gears. The phase shift φ may be of considerable influence on the stability of the trivial solution as demonstrated by the example (2).

The resulting matrices A and $B(t)$ in (16) possess the following structure:

$$
A = \begin{pmatrix}
0 & 0 & 0 & 1 & 0 & 0 \\
0 & 0 & 0 & 0 & 1 & 0 \\
0 & 0 & 0 & 0 & 0 & 1 \\
k_{120}^{(1)}r_1^2 + k_1^{(1)} & k_{120}^{(1)}r_1 r_2 & 0 & d_{12}^{(1)}r_1^2 + d_1^{(1)} & d_{12}^{(1)}r_1 r_2 & 0 \\
k_{120}^{(2)}r_1 r_2 & (k_{120}^{(2)} + k_{230}^{(2)})r_2^2 & -k_{230}^{(2)}r_2 r_3 & d_{12}^{(2)}r_1 r_2 & (d_{12}^{(2)} + d_{23}^{(2)})r_2^2 & -d_{23}^{(2)}r_2 r_3 \\
0 & -k_{230}^{(3)}r_2 r_3 & k_{230}^{(3)}r_3^2 + k_3^{(3)} & 0 & -d_{23}^{(3)}r_2 r_3 & d_{23}^{(3)}r_3^2 + d_3^{(3)}
\end{pmatrix}
$$

$$B = \begin{pmatrix} 0 & 0 & 0 & 0 & 0 & 0 \\ 0 & 0 & 0 & 0 & 0 & 0 \\ 0 & 0 & 0 & 0 & 0 & 0 \\ k_{121}^{(1)}\alpha r_1^2 & k_{121}^{(1)}\alpha r_1 r_2 & 0 & 0 & 0 & 0 \\ k_{121}^{(2)}\alpha r_1 r_2 & (k_{121}^{(2)}\alpha + k_{231}^{(2)}\beta)r_2^2 & -k_{231}^{(2)}\beta r_2 r_3 & 0 & 0 & 0 \\ 0 & -k_{231}^{(3)}\beta r_2 r_3 & k_{231}^{(3)}\beta r_2^2 & 0 & 0 & 0 \end{pmatrix}$$

with $\alpha := \sin(2\pi \cdot \omega \cdot t)$, $\beta := \sin(2\pi \cdot \omega \cdot t + \varphi)$, and superscript index (k) denoting the division of a constant by J_k, $k \in \{1, 2, 3\}$.

In the following figure, results of system (16) for $\varphi = 0$ and $\varphi = \pi$ are sketched in a parameter space which is formed by the coefficients k_{121}, k_{231} with $k_{121} = k_{231}$ of the oscillating terms in (18) and the frequency ω of teeth contacts.

For $\varphi = \pi$ points in the parameter space are marked by 'o' in the case of verified asymptotic stability, i.e., the stability test succeeded, and 'x' if stability could not be shown. The dotted and the dashed curve are approximations to the boundary of the domains of stability for $\varphi = 0$ and $\varphi = \pi$, respectively. Here, the domains of stability consist of all points between the ω-axis and the curves.

The influence of an increasing phase shift φ on the location of domains of stability is obvious:

- the critical frequency $\omega \approx 9000$ for $\varphi = 0$ has moved to $\omega \approx 7000$ for $\varphi = \pi$.
- the domains of stability have expanded for increasing values of φ.

It must be emphasized that points marked by 'o' are verified by the implemented algorithm, i.e., the boundary of the domain of stability must lie "above" this point for fixed value of ω. Possible reasons for the failure of the stability test at those points marked by 'x' are

- the accumulation of roundig errors which at any case are taken into account by the algorithm,
- the sufficient but not necessary character of the test, and
- the instability of the trivial solution.

Practical methods use the eigenfrequencies of the corresponding time-invariant system of (16) in order to estimate the location of "significant" domains of instability. Those algorithms by construction do not consider any phase shifts and, thus, yield results which may not really approximate the behavior of the time-dependent system. The observations made with the described mathematical model show that an expected relation between eigenfrequencies of the time-invariant system and the location of domains of instability of the time-dependent system is only valid for certain classes of problems which have not yet been specified.

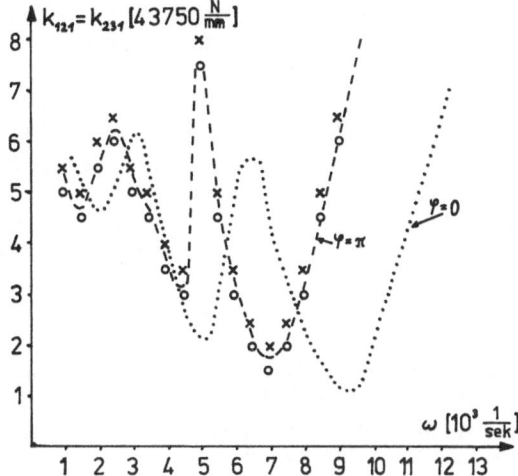

Figure 1. Results for system (16) with $\varphi = 0$ and $\varphi = \pi$.

6   Concluding Remarks

The application of the developed verifying algorithms to problems related to engineering modelling exhibits the practical advantages of the presented method:

- Due to the enclosure method numerical results are verified, and, thus, are reliable facts for the comparison of different models of simulation. The engineer may concentrate his work on the theoretical investigations of the problem instead of spending time in the error analysis of numerical algorithms. The not insignificant computational work is fully compensated by the achieved reliability of the quantitative results.

- Conventional methods may fail to recognize instability, and, thus, may produce misleading and erroneous decisions for modelling concepts.

- Systems of ordinary differential equations may be "typed in" without any preparatory work, except for a transfer to a non-dimensional version which is required anyway for efficient numerical work.

- Extensions to the problem of uncertain input data have partly been incorporated in the execution of enclosure methods, see [16].

The development of verifying algorithms and the implementation of problem solving routines yielding verified and enclosing bounds for the true solution must be emphasized. The concepts of parallelization can be applied efficiently to this type of methods and may establish a new generation of software tools and work packages which are time-efficient and yield validated results.

References

[1]   Alefeld,G./Herzberger,J.: An Introduction to Interval Computations, Academic Press, New York, 1983.

[2]   Arithmos Benutzerhandbuch, SIEMENS, Bestellnummer U2900-J-Z87-1, September 1986.

[3] Bleher,J./Kulisch,U./Metzger,M./Rump,S./Ullrich,Ch./Walter,W.: FORTRAN-SC: A Study of a FORTRAN Extension for Engineering / Scientific Computation with Access to ACRITH, Computing 39, November 1987 (p.93-110).

[4] Bohlender,G./Rall,L./Ullrich,Ch./Wolff von Gudenberg,J.: PASCAL-SC: Wirkungsvoll programmieren, kontrolliert rechnen, Bibliographisches Institut, Mannheim/Wien/Zürich, 1986.

[5] Cesari,L.: Asymptotic Behavior and Stability Problems in Ordinary Differential Equations, Springer Verlag, Berlin, 2nd edition, 1963.

[6] Cordes,D.: Verifizierter Stabilitätsnachweis für Lösungen von Systemen periodischer Differentialgleichungen auf dem Rechner mit Anwendungen, Universität Karlsruhe, Dissertation, 1987.

[7] Floquet,G.: Sur les équations différentielles linéaires à coefficients périodiques, Ann. Ecole Norm., Ser.2, 12, 1883.

[8] Hill,G.: On the part of the motion of the lunar perigee which is a function of the mean motions of the sun and the moon, Acta. Math. 8, 1886.

[9] IBM High Accuracy Arithmetic Subroutine Library (ACRITH), Program Description and User's Guide, SC 33-6164-02, 3rd edition, April 1986

[10] Kulisch,U./Miranker,W.: Computer Arithmetic in Theory and Practice, Academic Press, New York, 1981.

[11] La Salle,J./Lefschetz,S.: Die Stabilitätstheorie von Ljapunow, BI Hochschultaschenbücher, Band 194*, Bibliographisches Institut, Mannheim, 1967.

[12] Lohner,R.: Enclosing the Solutions of Ordinary Initial and Boudary Value Problems, published in: Computerarithmetic, Scientific Computation and Programming Languages, E.Kaucher/U.Kulisch/Ch.Ullrich(eds.), B.G.Teubner Verlag, Stuttgart, 1987.

[13] Mathieu,E.: Mémoire sur le mouvement vibratoire d'une membrane de forme elliptique, J. Pures Appl. 13, 1868.

[14] Moore,R.: Interval Analysis, Prentice Hall, Englewood Cliffs, N.J., 1966.

[15] Naab,K.: Stabilitätsuntersuchungen an linearen Systemen mit periodischen zeitveränderlichen Parametern, Fortschrittsberichte der VDI-Zeitschriften, Reihe Schwingungstechnik-Lärmbekämpfung, 11 Nr.44, VDI-Verlag, Düsseldorf, 1981.

[16] Spreuer,H./Adams,E./Holzmüller,A.: Enclosure of Output Distributions for Ordinary Differential Equations with Stochastic Input, published in : Computerarithmetic, Scientific Computation and Programming Languages, E.Kaucher/U.Kulisch/Ch.Ullrich(Editors), Teubner Verlag, Stuttgart, 1987.

Dietrich Cordes
Institut für Angewandte Mathematik
Universität Karlsruhe
Kaiserstrasse 12
D-7500 Karlsruhe
Federal Republic of Germany

Computing, Suppl. 6, 111–122 (1988)

The Periodic Solutions of the Oregonator and Verification of Results

E. Adams, A. Holzmüller, Karlsruhe, D. Straub, Neubiberg

Abstract — Zusammenfassung

The Periodic Solutions of the Oregonator and Verification of Results. The Oregonator is a numerically ill-conditional mathematical model in chemical kinetics involving nonlinear highly stiff ODEs. For the (stiffly coupled) "simplified Oregonator", periodic solutions in the phase plane can be confined to an annular closed strip S whose lateral extension can be made negligibly small within graphical accuracy. This result of the "Karlsruhe enclosure methods" has been obtained with a simultaneous verification of the existence of periodic solutions in S, making use of index theory and th Poincaré-Bendixson theory.

Die periodischen Lösungen des Oregonators und die Verifikation der Ergebnisse. Der Oregonator ist ein numerisch schlecht konditioniertes, mathematisches Modell in der chemischen Kinetik mit nichtlinearen, hochgradig steifen gewöhnlichen DGL. Für den (steif gekoppelten) „vereinfachten Oregonator" können periodische Lösungen in der Phasenebene auf einen ringförmigen Streifen S eingeschränkt werden. Die Breite dieses Streifens kann vernachlässigbar klein im Rahmen der zeichnerischen Genauigkeit gemacht werden. Verknüpft mit diesem Resultat der „Karlsruher Einschließungsmethoden" ist die Existenz periodischer Lösungen in S verifiziert worden, und zwar unter Verwendung der Index-Theorie und der Poincaré-Bendixson Theorie.

1. Introduction

The practical importance of enclosure methods can best be demonstrated by means of applications to selected nonlinear problems in analysis whose tough character has already been recognized by scientists or engineers. For this purpose, an evolution problem in chemical kinetics will be treated here. This domain is notorious since in realistic mathematical modelling pertinent systems of differential equations are highly nonlinear and stiff. The numerical complexities of this situation are enhanced when intermediates in the chain of chemical reactions oscillate. In the case of periodic oscillations, the boundary conditions of periodicity of the solution to be determined pose a further difficulty: the existence of the solution is to be verified. Additionally, the mathematical simulation of evolution problems in chemical kinetics is notoriously difficult due to significant uncertainties concerning

(i) the structure of the chemical reactions and hence of the model and

(ii) the values of parameters in a chosen model.

It is highly desirable to exclude uncertainties due to classical algorithms by means of enclosure methods. Reasons for discrepancies between computed and measured data can then be traced uniquely to the properties of classes (i) or (ii) of the mathematical model. The following translated quotation from Skrabal [8, p. VI] exhibits clearly that research chemists recognized numerical difficulties already many decades ago:

"Beside the technique of the experiment, there is the technique of numerical work. The latter is no less important than the former. Particularly in the numerical treatment of problems in (chemical) kinetics, again and again one encounters small (but) detrimental differences. These obstacles have to be circumvented. The pertinent techniques belong to the most important tools of kinetic computations; without their employment, the computation and its execution will invariably fail."

As a conclusion of this introduction, it should be observed that periodic chemical reactions are at the center of every process in biology. Even though they are of according practical and theoretical interest in modern science, their structure is hardly yet understood in biology, synergetics, etc. [7]. This is mainly due to shortcomings in mathematical modelling and reliable numerical methods.

2. On Oscillatory Chemical Reactions

Generally, a "net chemical reaction" with non-simple imput and output molecules does not proceed directly. Rather, there is a chain of reactions between intermediate molecules. When there exists more than one chain there is the possibility of an oscillatory process which follows different chains for different time intervals. According to [1, p. 102], the following additional conditions are necessary or "supportive" for the existence of oscillatory chemical reactions: the process (a) is far from any thermochemical equilibrium and (b) possesses at least one autocatalytic feedback. Oscillating *chemical* reactions possess well-defined periods only in exceptional cases and generally they are chaotic.

Approximately in 1970, R.M. Noyes, R.J. Field, and E. Körös began to search for chains of intermediates linking the input and the output of the oscillatory Belousov reaction. In 1972, they proposed the FKN-mechanism consisting of two coupled chains and 18 elementary reactions with 20 intermediates. They also chose a simplified mathematical model, the "Oregonator" [2], with five reactions and three intermediates to be called X, Y, and Z. Since 1975, a large number of papers have been published in chemistry or mathematics on oscillatory or periodic chemical reactions. Particularly the following references should be mentioned: the monograph [9] by J.J. Tyson and the proceedings [3] of an international conference at Aachen with 737 pages.

3. The Mathematical Model of the Oregonator

Field and Noyes investigated the reaction rates of the five reactions of their Oregonator model [2]. Accordingly, they derived a system of three nonlinear ordinary differential equations (ODEs) for the intermediates X, Y and Z, assuming an open system. In [2, p. 1879], the following nondimensional versions of these ODEs are derived:

(3.1)
$$\text{(i) } d\alpha/d\tau = s \cdot (\eta - \alpha\eta + \alpha - q\alpha^2),$$
$$\text{(ii) } d\eta/d\tau = s^{-1}(-\eta - \alpha\eta + f\rho),$$
$$\text{(iii) } d\rho/d\tau = w \cdot (\alpha - \rho).$$

The parameters q, s, and w depend in particular on certain reaction constants. The values of q, s, and w are not known precisely. According to [2] and [9], they can be confined to the following intervals:

(3.2) $s \in [63.2, 86.4]$, $w \in [0.177, 0.161]$, $q \in [0.536, 1.25] \cdot 10^{-5}$.

A sensitivity analysis of the Oregonator should take into account at least the values contained in these intervals. According to [2] and [9], there is no empirical information on f. In [2] f, s, q, and w are chosen so as to obtain a numerical approximation of a closed orbit:

(3.3) $q = 8.375 \cdot 10^{-6}$, $s = 77.27$, $w = 0.161$, $f = 1$.

The subsequent treatment of periodic solutions is confined to the choice of this "standard data".

The mathematical model of the Oregonator is

(a) uncertain due to the ambiguity of the input data and
(b) highly stiff because of the differences in the time constants s, s^{-1} and w of the ODEs in (3.1).

Consequently, (3.1 i) is singularly perturbed as compared with (3.1 ii) and (3.1 iii). For the qualitative analysis of problems of this kind, it s customary in literature to consider the associated asymptotic problem due to $s^{-1} \rightarrow 0$ in (3.1 i). In the natural sciences, this is said to be the "stiff coupling approximation". Accordingly (3.1) is replaced by the system

(3.4)
$$\text{(i) } \eta - \alpha\eta + \alpha - q\alpha^2 = 0,$$

(3.1 ii) and (3.1 iii).

Since chemical concentrations are nonnegative, only one of the two roots $\alpha = \alpha(\eta)$ of (3.4 i) is of interest. The mathematical model to be treated subsequently is chosen accordingly

(3.5)
$$\text{(i) } \alpha = \alpha(\eta) = (1 - \eta + \sqrt{4q\eta + (1 - \eta)^2})/(2q),$$
$$\text{(ii) } d\eta/d\tau = s^{-1}(-\eta - \alpha\eta + f\rho),$$
$$\text{(iii) } d\rho/d\tau = w \cdot (\alpha - \rho).$$

This model is said to be the simplified Oregonator.

4. Approximation of a Periodic Solution According to Field, Noyes, and Edelson

According to [2, p. 1879], the ODEs (3.1) "were integrated numerically by a Runge-Kutta started predictor-corrector technique". Numerical difficulties arose when α/η was smaller than 10^{-6}. Then, the employment of exceedingly small time steps could not prevent α and $d\alpha/d\tau$ to "oscillate wildly", due to rounding errors. This numerical "problem was eliminated by setting $d\alpha/d\tau = 0$ when α and $d\alpha/d\tau$ began to oscillate during the integration". In [2], then $\alpha = \alpha(\eta)$ was calculated by use of (3.5i). "When η fell to the point where the equations were numerically well behaved again, the constraint due to (3.5i) was lifted." A spot check by use of a 'Gear integration method' indicated that the alternating employment of (3.1) and (3.5) "does not lead to appreciable error in the analysis of the Oregonator" [2, p. 1880].

According to [2, p. 1881], "when integrations were carried out for more than one cycle, successive trajectories followed each other to within 0.1 precision of the calculation." In [2], additionally, the approximation of the cycle was determined by use only of (3.5). According to [2, p. 1882], "the results ... indicate that this is a reasonable approach throughout much of the cycle".

A more or less quick approach of the cycle was observed for trajectories starting outside the cycle. The numerical results in [2] exhibit the following features:

(i)   within the entire τ-period of approximately 300, there are *relatively* small rates of change of η and ρ and particularly α,
(ii)   which, however, are interrupted by "peaks" covering a τ-interval of width less than five such that
(iii) α first increases and then decreases both by five decades,
(iv) η first decreases by two decades and then increases by five decades, and
(v) ρ increases by five decades.

As compared with approximations presented in [2], pertinent experimental results exhibit considerably smaller rates of change of certain concentrations, of only two to three decades [9, p. 30]. The reasons for these discrepancies cannot be traced back uniquely to either the model or the employed numerical technique or the experiment. In fact, in [2], there is no estimate of the influence of the rounding or the procedural errors. This situation suggests the employment of enclosure methods to eliminate the uncertainties of the computed results for a given method.

5. The Existence of Periodic Solutions

The following "2×2 system" is considered

$$(5.1) \qquad\qquad x' = X(x), \qquad x = x(t);$$

X is a real continuous vector function defined on a bounded open subset D of the real plane such that for each real t_0 and each point $x_0 \in D$ there exists a unique vector solution $\varphi = \varphi(t; x_0)$ of (5.1) satisfying $\varphi_0(t_0; x_0) = x_0$.

The set of all points $P(t)$ of D with coordinates $(\varphi_1(t),\ \varphi_2(t))$ for $t_0 \leq t < +\infty$ is called a positive half-path of the system (5.1). Then there holds according to [5, p. 321]:

Theorem (Poincaré-Bendixson): *Let R be a closed bounded region consisting of nonsingular points of a 2×2 system $x' = X(x)$ such that some positive halfpath H of the system lies entirely within R. Then either H is itself a closed path, or it approaches a closed path, or it terminates at a stationary point.*

In the case of a system of two autonomous explicit ODEs of the first order each, index theory establishes relationships between stationary points and periodic orbits in $\mathbb{R}^2$. According to [4, p. 151].

      (*i*) inside every closed orbit there must be at least one stationary point.

(5.2) (*ii*) if there is only one stationary point inside a closed orbit
          then it cannot be a saddle point.

6. Verification of the Existence of Periodic Solutions of the Simplified Oregonator

In this section it will be shown how the general theorem of Poincaré-Bendixson can be applied by use of an appropriate computer to prove the existence of periodic solutions of the (autonomous) differential equations (5.1). This will be presented for the case of the simplified Oregonator (3.5).

Equating the right hand sides of (3.5ii) and (3.5iii) to zero it is seen that (η_{01}, ρ_{01}) is the only stationary point in the (pertinent) first quadrant, with the following values by use of the standard data from [2]:

(6.1) $\eta_{01} \in 0.99795778 + [0, 1]\, 10^{-8}, \qquad \rho_{01} \in 0.48867803 \cdot 10^{+3} + [-1, 1] \cdot 10^{-5}.$

The next step in the investigation of the existence of periodic solutions consists in the derivation of the linear variational system of (3.5) with linearization at the stationary point $(\eta_{01},\ \rho_{01})$. This yields a system of two linear ODEs with constant coefficients, whose eigenvalues are the roots of the characteristic equation

(6.2) $\lambda^2 + P\lambda + R = 0,$ where $\begin{aligned}P &= P(\eta_{01}, \rho_{01}) \approx -6.15,\\ R &= R(\eta_{01}, \rho_{01}) \approx\ \ 4.17 \cdot 10^{-3}.\end{aligned}$

By use of an obvious notation for intervals, the solutions of (6.2) are:

(6.3) $\lambda_1 \in 1.0187_6^8 \cdot 10^3, \qquad \lambda_2 \in 1.3_{29}^{36} \cdot 10^{-3}.$

Since both eigenvalues are positive the stationary point (6.1) is a source. This test shows, that it makes sense to continue the search for a periodic orbit in the first quadrant. In the case of opposite signs of the eigenvalues λ_1 and λ_2 a periodic orbit does not exist in the first quadrant (cf. (5.2)).

Remark: In many computed examples small changes of η_{01} nad ρ_{01} yielded $\lambda_1 \cdot \lambda_2 < 0$, i.e. the stationary point was (or seemed to be) a saddle point. These changes of η_{01} and ρ_{01} might be caused by small changes of f or q.

E. Adams et al.:

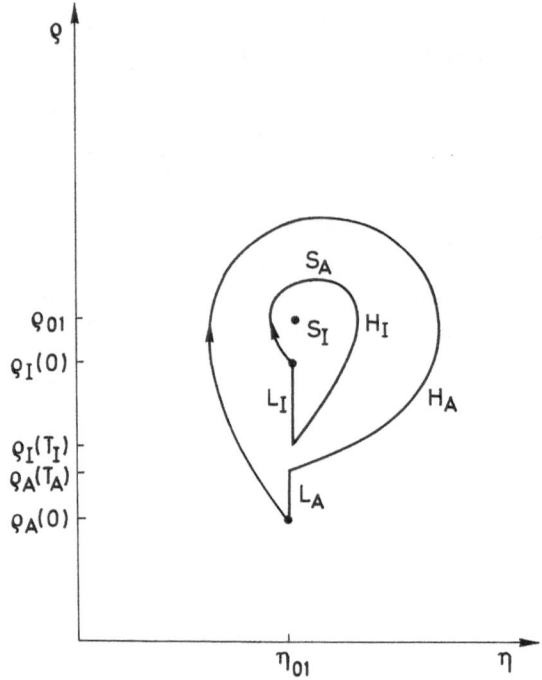

Figure 6.7. Schematic representation of sets employed in section 6

The idea of proving the existence of a periodic orbit and of computing sharp bounds for this orbit is as follows:

Suppose two solutions $A(t)=(\eta_A,\ \rho_A)^T(t)$ and $I(t)=(\eta_I,\ \rho_I)^T(t)$ of (3.5), $t\geq 0$ are known, which are subject to the initial conditions

(6.4) $$\eta_A(0)=\eta_I(0)=\eta_{01},\qquad 0<\rho_A(0)<\rho_I(0)<\rho_{01}.$$

Suppose further that there exist $A(T_A)$ and $I(T_I)$ such that

(6.5) $$0<\rho_A(0)<\rho_A(T_A)<\rho_I(T_I)<\rho_I(0)<\rho_{01},$$
$$\eta_A(T_A)=\eta_I(T_I)=\eta_{01}$$

For the sake of clarity we assume $T_A=\min\{t\,|\,t>0,\ \eta_A(t)=\eta_{01},\ \rho_A(t)<\rho_{01}\}$, $T_I:=\min\{t\,|\,t>0,\ \eta_I(t)=\eta_{01},\ \rho_I(t)<\rho_{01}\}$. Let S_A be the region bounded by $H_A:=\{A(t)\,|\,t\in[0,\ T_A]\}$ and the line segment $L_A:=\{(\eta_{01},\ \rho)\,|\,\rho_A(0)\leq\rho\leq\rho_A(T_A)\}$ and S_I the region bounded by $H_I:=\{I(t)\,|\,t\in[0,\ T_I]\}$ and the line segment $L_I:=\{(\eta_{01},\ \rho)\,|\,\rho_I(0)\geq\rho\geq\rho_I(T_I)\}$. Finally suppose

(6.6) $$(\eta_{01},\rho_{01})\in\mathrm{int}(S_A)\cap\mathrm{int}(S_I),$$

where $\mathrm{int}(M)$ denotes the interior of a set M. These assumptions simply mean, that A spirals inward (towards the stationary point $(\eta_{01},\ \rho_{01})$) and I spirals outward. By the specific choice of T_A and T_I there is only one revolution of A and I about the stationary point (cf. Figure 6.7).

Then there holds:

(6.7) **Theorem:** *If the conditions* (6.4)−(6.6) *are satisfied and if* $\dfrac{d\eta}{d\tau}$ $(\eta,\ \rho)<0$

for all $(\eta,\ \rho)\in L_A\cup L_I$ *then there exists a periodic solution of* (3.5) *in* $S:=(S_A\backslash S_I)\cup H_I$.

Proof: (Outline) By construction S is a closed and bounded region consisting of nonsingular points of (3.5). Every half-path H starting in S cannot leave S since it cannot cross H_I or H_A due to the fact, that every initial vector $(\eta_0,\ \rho_0)^T\in S$ defines a unique solution of the corresponding IVP (3.5). Also H cannot cross L_I or L_A since $\dfrac{d\eta}{d\tau}$ $(\eta,\ \rho)<0$ for all $(\eta,\ \rho)\in L_A\cup L_I$. Thus S and any H starting in $\mathrm{int}(S)$ satisfy the assumptions of the theorem of Poincaré-Bendixson (cf. section 5).

Furthermore S is a closed set which does not contain any stationary point. Hence either H is itself a periodic orbit (in S) or it approaches a closed path, which therefore has to be in S, too. □

Remarks: (1) Several of the conditions were stated for the purpose of a simplification of the proof of theorem 6.7. They can easily be relaxed.
(2) The existence of one and only one periodic solution in S cannot be shown by use of the employed methods.
(3) The theorem is easily generalized to a wide class of autonomous 2×2 systems. Basically only the employed line segments and the condition "$d\eta/d\tau\ (\eta,\ \rho)<0$" have to be replaced by similar conditions, ensuring that a halfpath cannot leave the region bounded by H_A and H_I and the corresponding line segments.

Approximations of the solutions A and I are obviously not sufficient to guarantee that condition (6.5) of Theorem 6.8 is satisfied. Consequently, enclosures $[A(\tau)]:=[\underline{A}(\tau),\ \bar{A}(\tau)]$ of $A=A(\tau)$ and $[I(\tau)]:=[\underline{I}(\tau),\ \bar{I}(\tau)]$ of $I=I(\tau)$ were determined. Condition (6.5) is satisfied provided this is so for the following inequality:

(6.8)
$$0<\rho_A(0)<\underline{\rho}_A(T_A)<\bar{\rho}_I(T_I)<\rho_I(0)<\rho_{01},$$
$$\eta_A(T_A)=\eta_I(T_I)=\eta_{01}.$$

7. Numerical Experience Gained in the Construction of $[A](\tau)$ and $[I](\tau)$

According to [6], the employed enclosure method for initial value problems rests on a suitable extension of an explicit one-step difference method of order p and with time-step size h. An inspection of results from [2] revealed the necessity of a suitable control of the artificial parameters h (and p) of the enclosure algorithm. The choice of an efficient control principle is generally difficult in the determination of approximations of solutions of initial value problems; in fact, estimations of the *local errors* then are usually the only practical approach. In the case of enclosure methods, the computed componentwise distances of the upper and the lower bounds provide a direct measure of the *global errors*.

According to numerical experience, generally there is an overlinear self-amplification tendency of the widths of the enclosures as a time-like variable increases. If uncontrolled, this will eventually lead to an exponent overflow. In the construction of $[A(\tau)]$ and $[I(\tau)]$, a control was executed as follows by the user:

(i) upon detecting a growth tendency of the enclosures, h was tentatively decreased (or p increased);
(ii) the opposite change(s) were carried out occasionally in the absence of a growth tendency.

In each one of the constructions of $[A(\tau)]$ and $[I(\tau)]$ for one full revolution, these controls were carried out approximately 15 times. In view of narrow enclosures and the total computation time, it was more effective to change h than p.

Upon changing h (or p) at a time τ_{ch}, it was always possible to restart the construction of $[A]$ at *one* point $p_A \in [A(\tau_{ch})]$. According to the direction field and all solutions starting in $[A(\tau_{ch})]$, the "outermost" solution is chosen for the continuation for $\tau > \tau_{ch}$. Correspondingly, the "innermost" solution is chosen in the case of $[I(\tau)]$.

An additional control was employed for a transfer between (3.5i) and the (mathematically equivalent) representation

$$(7.1) \qquad \alpha = \alpha(\eta) = 4q\eta/((2q)(1-\eta-\sqrt{4q\eta+(\eta-1)^2})).$$

In view of rounding errors, (3.5i) was used when $0 < \eta \ll 1$ and (7.1) was used when $\eta \gg 1$.

Remark: This control will be automated for future applications.

8. Numerical Results

The following quantitative results concerning the enclosure $[I]$ are presented in *Table 8.1*:*)

(i) selected values τ_i of the time τ which are indicated in Figure 8.1,
(ii) the local value of h employed at $\tau = \tau_i$, and
(iii) the total number of time steps for $\tau \in [0, \tau_i]$.

The results for τ_i and h confirm the need for a suitable control of h (and p). The peaks referred to in [2] are represented by the very rapid "motion past the orbit" for $\tau \in [0, \tau_3]$.

Table 8.2 presents a comparison of the computed enclosures $[A]$ and $[I]$ at selected points "1", "2", "3", and "4" which are indicated in Figure 8.1. The

---

*) All numerical results were obtained by use of a kws-EB 68/20 PC supporting the computer language PASCAL-SC.

Table 8.1:
Results for the Enclosure $[I]$ of the Inner Solution

Time τ_i	Total Number of Executed Time Steps	Step Size h
$\tau_0 = \quad 0$	0	$5 \cdot 10^{-5}$
$\tau_1 = \quad 10^{-2}$	200	$5 \cdot 10^{-5}$
$\tau_2 = \quad 2.448$	8700	10^{-5}
$\tau_3 = \quad 6.069$	11520	$5 \cdot 10^{-3}$
$\tau_4 = \quad 83.469$	12741	10^{0}
$\tau_5 = 258.469$	12916	10^{0}
$\tau_6 = 300.35$	14271	10^{-4}

Table 8.2:
Results in the Phase Plane for the Enclosure $[A]$ of the Outer Solution
and $[I]$ of the Inner Solution

Point	$[I]$ with Coordinates: $[\eta]$	$[\rho]$	$[A]$ with Coordinates: $[\eta]$	$[\rho]$
"1"	199	27565	199	27567
"2"	1703	3353	1703	3354
"3"	109	1.0081	109	1.0079
"4" at time $\tau = 0$	1	20	1	1
"4" at time τ of one full revolution	1	[5.378, 5.384]	1	[4.80, 5.85]

Table 8.3:
Values of ρ on the Line $\eta = 1$ at the Beginning and at the End of One
Full Revolution in the Phase Plane

$\rho_A(0)$	$\underline{\rho}_A(T_A)$	$\underline{\rho}_I(T_I) \approx \bar{\rho}_I(T_I)$	$\bar{\rho}_A(T_A)$	$\rho_I(0)$
1	4.8	5.4	5.85	20

orbits presented in Figure 8.1 and the corresponding graph in [2] coincide within graphical accuracy. With the exception of point "4", only the coincident leading mantissa digits of the upper and the lower bounds are presented in the case of the coordinates of either $[A]$ or $[I]$. The non-coincidence at point "4" is due to the fact that the foreseeable completion of one full revolution made it unnecessary to control h (and p) during the terminal stage. With reference to the inequalities (6.5), *Table 8.3* exhibits for $\eta = \eta_{01} \approx 1$

(i) the chosen initial values $\rho_A(0)$ and $\rho_I(0)$,
(ii) the computed bounds $\underline{\rho}_A(T_A)$ and $\bar{\rho}_A(T_A)$, and
(iii) the correspondingly computed bounds $\underline{\rho}_I(T_I)$ and $\bar{\rho}_I(T_I)$.

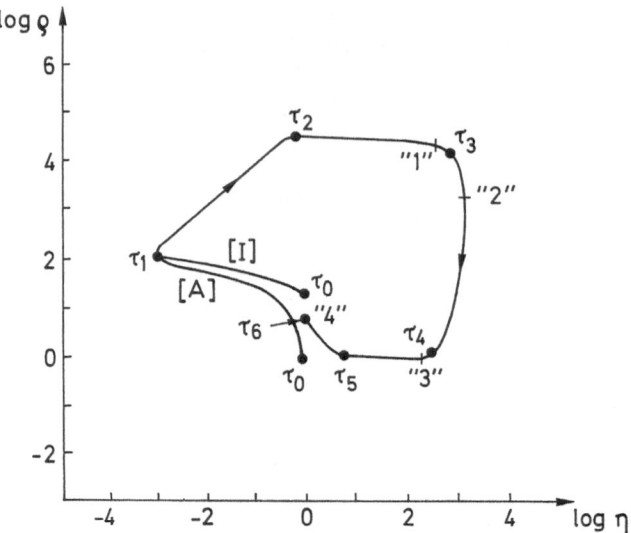

Figure 8.1. Enclosures [A] of the Outer Solution A and [I] of the Inner Solution I in the Phase Plane. See Table 8.1 for $\tau_1, \ldots, \tau_6$ and Table 8.2 for "1", ..., "4"

A print-out of computed results shows that [A] and [I] are overlapping for certain time-intervals between the points "1" and "3". A continuation of the construction of $[A(\tau)]$ and $[I(\tau)]$ for $\tau > \mathrm{Min}\{T_A, T_I\}$ yields a "strip" containing a periodic solution whose lateral width can be made exceedingly small as compared with $\sqrt{\eta_{01}^2 + \rho_{01}^2}$ which is the only relevant chemical entity (this width is thus smaller than graphical accuracy).

9. Concluding Remarks

9.1 Mathematical and Numerical Aspects

This paper confirms that the "Karlsruhe enclosure methods" are of practical use even in the case of classes of large and "non-nice" problems from fields outside of mathematics with ODEs involved. In fact, the Oregonator has always been known as a notoriously tough model. A combination of index theory and Poincaré-Bendixson theory establishes the existence of periodic solutions in an annular set S whose lateral width can be made negligibly small within graphical accuracy. The computability of this enclosure (in the presence of the unavoidable interval overestimates) verifies some kind of practical stability with respect to computational errors. The *combined* execution of all these qualitative and quantitative tasks is obviously more efficient as compared with separate treatments.

9.2 Relevance of the Karlsruhe Enclosure Methods in Mathematical Simulation

Oscillatory solutions of large nonlinear systems in dynamics are characterized by the occurrence of

(i) a wide band of frequencies,
(ii) the beat phenomenon, and/or
(iii) sharp peaks.

Mathematical models with these features of their solutions are common in contemporary chemical kinetics, physics, biology, (celestial) mechanics, etc. These features demand an efficient automatic control of the artificial parameters of the employed numerical method. In the case of enclosure methods, this can be based on the actual global error. Consequently, the enclosure methods employed here possess a built-in self-guidance property. This "safety-device" enables a user to employ step sizes h which almost always are (considerably) larger than the ones which are customarily applied in the case of approximation methods.

The solutions of the Oregonator are exceedingly sensitive with respect to small parameter variations. A results of this kind is usually interpreted to demand the execution of a suitable structural stability analysis of the model. Consequently, the Oregonator should be replaced tentatively by less simplified versions of the FKN-mechanism, with more than only three intermediates. The total reliability of the results of enclosure methods then enables a user to concentrate his/her efforts on these simulation aspects.

References

[1] Eppstein, I. R., et al.: Oszillierende chemische Reaktionen, Spektrum der Wissenschaften, Mai 1983, p. 98 – 107.
[2] Field, R. J., Noyes, R. M.: Oscillations in chemical systems, IV. Limit cycle behavior in a model of a real chemical reaction, The Journal of Chemical Physics, vol. 60, 1974, p. 1877 – 1884.
[3] Franck, U. F., et al.: Kinetics of Physicochemical Oscillations, Preprints of Submitted papers for Discussion Meeting held by Deutsche Bunsengesellschaft für Physikalische Chemie at Aachen, September, 1979.
[4] Guckenheimer, J., Holmes, P.: Nonlinear Oscillations, Dynamical Systems, and Bifurcations of Vector Fields, Springer-Verlag, New York, 1983.
[5] Jordan, D. W., Smith, P.: Nonlinear Ordinary Differential Equations, Clarendon Press, Oxford, 1977.
[6] Lohner, R. J.: Enclosing the Solutions of Ordinary Initial and Boundary Value Problems, p. 255–286 in: Computerarithmetic, Editors: E. Kaucher, U. Kulisch, Ch. Ullrich, B. G. Teubner, Stuttgart, 1987.
[7] Nicolis, G., Prigogine, I.: Self-Organization and Nonequilibrium Systems, J. Wiley & Sons, New York, 1977.
[8] Skrabal, A.: Homogenkinetik, Experiment und rechnerische Grundlagen, Verlag Th. Steinkopf, Dresden, 1941.
[9] Tyson, J. J.: The Belousov-Znabotinskii Reaction, Springer-Verlag, Berlin, 1976.

Dr. E. Adams, Dr. A. Holzmüller
Institut für Angewandte Mathematik
Universität Karlsruhe
Kaiserstrasse 12
D-7500 Karlsruhe
Federal Republic of Germany

Dr. D. Straub
Institut für Thermodynamik
Fakultät für Luft- und Raumfahrttechnik
Universität der Bundeswehr München
Werner-Heisenberg-Weg 39
D-8014 Neubiberg
Federal Republic of Germany

Computing, Suppl. 6, 123–136 (1988)

On Arithmetical Problems of Geometric Algorithms in the Plane[1]

Th. Ottmann, G. Thiemt, Ch. Ullrich, Freiburg i.B.

Abstract — Zusammenfassung

On Arithmetical Problems of Geometric Algorithms in the Plane. In the last years computational geometry has achieved a mature status within the framework of algorithms and data structures. Hundreds of new solutions for geometric problems are avaiable now. Corresponding algorithms are distinguished by clever data structures, rigorous proofs and nontrivial complexity analysis.

The application of the new algorithms in real systems is less successful. One of the reasons for this phenomenon is given by the numerical problems occurring during execution of nearly all geometric algorithms.

This paper gives an introduction to the latter class of problems in geometric algorithms. The scan–line algorithm for computing all pairs of intersecting line segments in the plane serves as model example for the isolation of basic operations which have to be handled numerically correct. It can be shown that the optimal evaluation of arithmetic expressions provides a solid tool for the solution of the problems.

Über arithmetische Probleme bei geometrischen Algorithmen in der Ebene. In den letzten Jahren hat die algorithmische Geometrie einen hohen Standard im Hinblick auf Algorithmen und Datenstrukturen erreicht. Hunderte neuer Lösungen für geometrische Probleme stehen heute zur Verfügung. Die zugehörigen Algorithmen zeichnen sich durch ausgeklügelte Datenstrukturen, strenge Beweise und eine nichttriviale Komplexitätsanalyse aus.

Die Anwendung der neuen Methoden in realen Systemen ist weniger erfolgreich. Einen der Gründe für dieses Phänomen bilden die numerischen Schwierigkeiten, welche bei der Ausführung nahezu aller geometrischen Algorithmen auftreten.

Die vorliegende Arbeit gibt eine Einführung in diese Problematik. Als ein Modell für die Isolierung der Elementaroperationen, welche numerisch korrekt zu behandeln sind, dient der scan-line Algorithmus zur Berechnung aller Paare sich schneidender Liniensegmente einer vorgegebenen Segmentmenge in der Ebene. Es zeigt sich, daß die optimale Auswertung arithmetischer Ausdrücke ein solides Werkzeug zur Lösung der Probleme darstellt.

1   Introduction

Geometry is certainly one of the oldest areas of mathematics with roots already in ancient times. However, algorithmic and combinatorial aspects of geometry

---

[1] The work of the first two authors was carried out under DFG Grant Ot64/5–1. A preliminary version has been presented at the 3rd ACM Symposium on Computational Geometry, Waterloo, 1987.

and questions of how to solve a geometric problem by using computers have attracted growing interest only just recently. The main reason is, of course, that changing demands emanate from new applications such as computer graphics, picture processing, geography, carthography, robotics, VLSI–design, and many other areas. Roughly speaking, at the very beginning of the electronic data processing age the processing of numeric data in batch mode predominated, while nowadays the processing and management of spatial, i. e. geometric, data prevails. Thus, within the last ten years a new and fascinating research area has grown up: computational geometry. It is now a well established area within the framework of algorithms and data structures and has obtained a quite mature status.

The notion "computational geometry" as we understand it in this paper was introduced by M. Shamos who wrote a thesis with this title some ten years ago. This thesis appeared in 1985 in heavily revised and much enlarged form as the first textbook on the field, cf. [PS]. In the meantime the field has blossomed considerably. The current trends can be best observed by a look into the proceedings volumes of the annual ACM Sigact/Sigraph conference on Computational Geometry.

What are the problems dealt within this area? We give a random but typical and incomplete list of geometrical problems:

- Convexity: Given a set S of n points in d–space, compute its convex hull.

- Intersection and visibility: Given a set of line segments, rectangles, or polygons in the plane. Report all intersecting pairs; compute the visible objects, if the given set of objects is a 2–dim. projection of a 3–dim. scene.

- Compute for a set of geometric objects the area, perimeter, contour, connected components, certain closures, etc.

- Decompose (partition) a set of polygons into triangles, a set of rectilinear polygons into quadrilaterals etc.

- Point location: Given a planar subdivision of the plane, e. g. a triangulation. Determine for a given point the region into which the point falls.

Literally hundreds of new solutions for geometric problems of this kind are now available. Not only *new* data structures like segment–, interval–, and priority search trees, have been invented, but also *classical* structures, like Voronoi-diagrams, have been revived. Powerful design principles like the scan–line paradigm, geometric divide–and–conquer, and geometric transformations were successfully applied many times. They should certainly belong to the toolbox of theorists and practitioners alike nowadays working in the field. Thus, from a theoretical point of view computational geometry was quite successful. It is distinguished by a large number of sophisticated algorithms, clever data structures, rigorous proofs, nontrivial complexity analysis, and a number of challenging open problems. These are enough reasons for an area of theoretical research to be still blossoming.

From a more practical point of view computational geometry was not similarly successful. The application of new algorithms in real systems is still exceptional. Many reasons for this phenomenon are quite well known. We mention a few of them. Theoretical research is still too much concentrated on asymptotic worst case efficiency. (In practice it is much more important to achieve a good average behavior for sets of real data.) Other important and still unresolved questions concern the problem of representing and manipulating complex three–dimensional objects, the problem of how to treat and not to exclude the numerous "special cases" occurring in practice, the development and implementation of powerful *standard* structures which can be used in many algorithms instead of a large number of unique, application–dependent structures. Last but not least the *numerical problems* involved in geometric computations have to be solved.

This paper deals with the latter problem. There are only a very few papers in the open literature which address the numerical problems in geometric computations. The textbook mentioned above does not report anything about these problems. That the numerical problems are nevertheless important from a-practical point of view has been emphazied in [F]. We refer to two papers which represent two extreme approaches in solving the numerical problems in geometric computations.

Beretta's Ph.D. thesis [Ber] contains a section on the influence of finite arithmetic on the implementation of the scan–line algorithm. He derives the following requirement for the floating–point arithmetic: Assume that the input data have integer coordinate values in the range $0..M$. Then at least $5 \log M + 4$ bits are required for the storage of the fraction part of floating–point numbers in order for the algorithms to run correctly. In summary, Beretta relies largely on the floating–point arithmetic available on a computer; however, his approach of solving the numerical problems in geometric computations is rather limited and fairly ad hoc.

The other extreme may be represented by V. Akman's thesis [A]: It contains several algorithms for computing shortest paths in 3-space which avoid a given set of polyhedra. The algorithms are implemented in Lisp and use functions from Macsyma, a very sophisticated computer algebra system, cf. [M]. The aim is to built up a Macsyma-like geometer's workbench which provides its user with a large number of symbolic geometric operations.

Our approach lies in between these two extremes. We require more than just standard floating–point arithmetic, but we do not rely on the ability of a computer to carry out "arbitrary" computations occuring in geometric algorithms exactly. We show how to achieve numerical stability for a large class of geometrical algorithms.

Whether or not the method described in this paper achieves these goals will be discussed in the final Section 5. Of course, it has to be confirmed by real implementations. In Section 2 we give several examples of arithmetical problems occuring in geometric computations. As a running example we choose the well known scan–line algorithm for reporting all pairs of intersecting line segments in the plane, cf. [BO]. We will illustrate a possible solution of the numerical problems for this example in Section 3. Section 4 reports some experimental results obtained on a computer whose software supports the exact computation of scalar–products.

2   Arithmetical Problems in Geometric Computations

The implementation of a geometric algorithm on a given machine not only requires to realize more or less sophisticated data structures, which are usually not available as standard structures in common programming languages. The implementor has also to take into account numerous "special" or "extreme" cases and the fact that computers do not always yield exact and correct results, even if the input data are assumed not to be affected with uncertainty.

In order to be more precise let us consider the following example: A basic problem in computer graphics is the elimination of hidden lines in a two–dimensional representention of three–dimensional objects ("hidden–line–elimi-nation–problem"). If the object is modelled by rectilinear edges in the three–dimensional space, the projection of these edges into the plane yields a set of line segments. The elimination of the hidden parts of the line segments comprises as an important subtask the computation of all existing intersection points which can be done by the well known scan–line algorithm [BO]. This algorithm computes all pairs of intersecting line segments for a given set of line segments in the plane. Each line segment is assumed to be given as the pair of its endpoints, each endpoint as the pair of its (x, y)–coordinates. We assume that the given input is exact, i.e. the coordinate values are machine numbers in the floating–point system of the given machine. Thus, *original points are exact*. In order to compute the intersecting pairs a vertical scan–line is swept from left to right through the set of input data in discrete steps (see Figure 1). The sweep of the scan–line is controlled by the *event queue*, i.e. the scan–line halts at each point in this queue. The event queue is initialised as the set of left and right endpoints of the given segments in increasing x–order; entries with equal x–values are ordered according to their y–coordinates. Remaining ties are broken arbitrarily. During the scan detected intersection points have to be inserted into the event queue according to their (exact!) x–coordinate. At each stage the active line segments (which are currently cut by the scan–line) are maintained in a dynamic *vertical structure L*. *L* stores the active line segments according to increasing y–values of their intersections with the scan–line. In order to correctly maintain this y–order intersecting line segments have to be swapped at their intersection point (see Figure 2).

Let us analyse the operations occurring in this algorithm from a numerical point of view: A new line segment has to be *inserted* into the vertical structure L whenever a new segment begins. That means, we have to determine the position

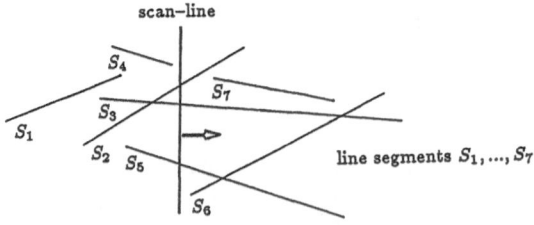

Figure 1. Line segments and scan–line

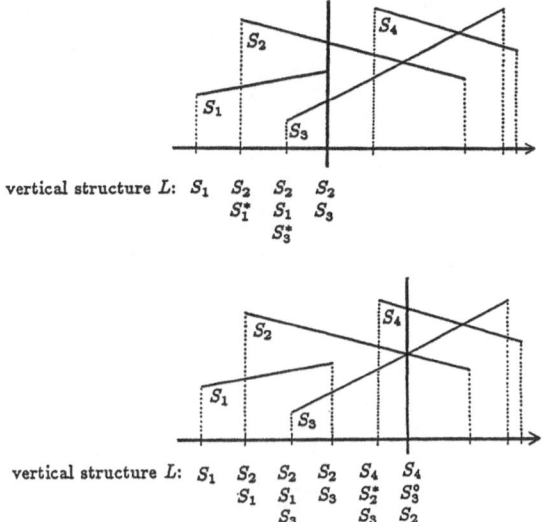

vertical structure L: S_1 S_2 S_2 S_2
S_1^* S_1 S_3
S_3^*

vertical structure L: S_1 S_2 S_2 S_2 S_4 S_4
S_1 S_1 S_3 S_2^* S_3^o
S_3 S_3 S_2

* means intersection test without success.
o means succesful intersection test; the intersection
  x–coordinate $S_2 \cap S_3$ is inserted into the event queue.
+ means successful intersection test; the intersection
  x–coordinate of $S_3 \cap S_4$ is inserted into the event queue.

Figure 2. Changes in the vertical structure L during execution of the scan–line algorithm

of its left endpoint in the currently valid y–order of segments in L. That requires to perform several *point–line–location tests*.

- For a given original point p and a line segment given by a pair of two original endpoints, determine whether p lies on, above or below the line.

The *deletion* of a line segment from L and the *swapping* of two segments in L do not require numerical operations: By using a name–tree it's possible to access the segments directly. Thus, from a numerical point of view, the point–line–location test is the only crucial operation on L.

The critical operation on the event queue is to *determine the x–order* between computed intersection points and/or original points. We must be able to correctly locate the position of a computed intersection point in the x–order of elements in the event queue. Observe that the event queue may not only contain original points but also some (already previously) computed intersection points. The following two examples describe the possible errors during the scan–line algorithm and the consequences for the vertical structure.

Example 1: Three segments A, B, C are intersecting within a small area. Because of rounding errors the computation of the intersection point $B \cap C$ yields the smallest value for the x–coordinate leading to an incorrect vertical structure (see Figure 3).

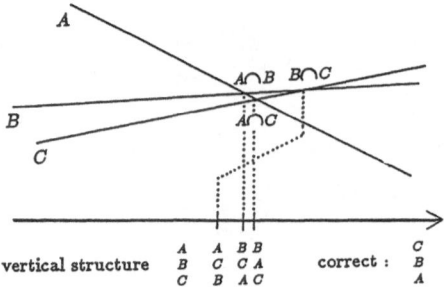

Figure 3. Change of order in the event queue because of erroneous computation and its consequences for the vertical structure

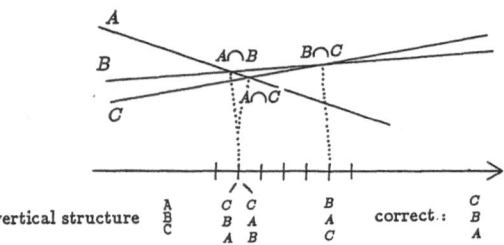

Figure 4. Change of order in the event queue because of rounding to the same machine number and its consequences for the vertical structure

Example 2: Here two of three segments A, B, C are intersecting within a small area. The computation of the intersection points $A \cap B$ and $A \cap C$ yields the same machine number for the x–coordinate. By accident the intersection point $A \cap C$ is handled at first (See Figure 4).

Determining the correct x–order must be possible even if the computed intersection points are not machine numbers in the floating–point number system on the given computer. Relying on the available floating–point number system may lead to completely wrong results for the critical operations in the scan–line algorithm mentioned above. We demonstrate this by an example for the point–line–location test.

Let a line $f(x, y) = 0$ be given, where

$$f(x, y) = 1934.5x - 656.38y - 2.9236.$$

We want to locate the position of points p_1, p_2, p_3 in a decimal floating–point number system with 5 significant decimal digits.

point	p_1	p_2	p_3
location with respect to line f	(2269.2, 6687.8)	(2269.3, 6688.1)	(0.023628, 0.025095)
correct:	on	above	below
computed:	below	below	on

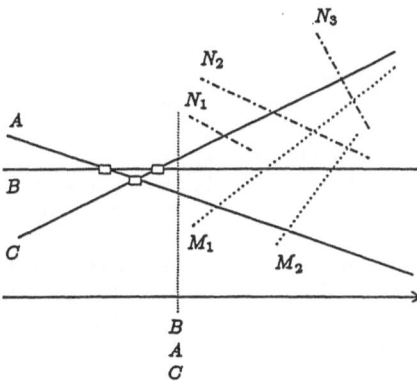

Figure 5. Change of intersections is assumed leading to the given vertical structure

Already one wrong answer to a point–line–location test or one wrong x–order determination for a computed point in the event queue may have a disasterous global effect for the scan–line algorithm: It may happen that many true intersections remain undetected, as Figure 5 demonstrates.

The segments N_1, N_2, N_3 are located above segment B and, therefore, are inserted at the top of the vertical structure. On the other hand the segments M_1, M_2 are inserted at the bottom of the vertical structure. Since the order of the segments B, A, C will not be changed by further execution of the scan–line algorithm, the intersection points of the M– and N–segments are not computed.

Point–line–location tests and determining the x–order between computed and/or original points are just two examples of operations occurring in many geometric algorithms. We mention some others which are of a similar nature:

- *Identity test* for two points p and q, where p and q are original points or points computed from original points in a finite number of steps. (Example: p and q are intersection points of two lines given by two pairs of original points p_1, p_2 and q_1, q_2. Of course, if the intersection points are not machine numbers we have to represent them in some way!)

- *Test* whether two lines given by two pairs of orignial or computed points are *parallel*.

- *Circle inclusion test:* Given three original points and a computed intersection point, determine whether or not the computed point is inside the circle given by the three original points. (This operation occurs in the divide–and–conquer algorithm for computing the Voronoi–diagram, cf. [PS], [GS].)

These are examples of operations of the *test–type*. The result of such an operation is a value from a small finite set of different values. Test operations should always have correct results.

Of course, we also want to compute *geometric results*, i. e. geometric objects as the output of a geometric algorithm, like points, lines, circles, triangles etc. defined

Table 1

geometric operation	applications	representation of the geom. objects
A) Test–operations		
1) (x, y)–order for two points	searching problems, intersection problems, Voronoi diagrams	a) two exact points b) one exact point one intersection point c) two intersection points
2) point–line–location test	convex hulls, point–location, intersection problems	a) one exact point, line b) one intersection point, one line
3) orientation of an angle	convex hulls	a) three exact points
4) intersection test	point–location, intersection problems	a) two segments b) two lines c) one line, one segment
5) incircle–test (testing, whether a point lies in a circle, given by three points on the boundary)	Voronoi diagrams	a) four exact points
6) comparison of two gradients	kernel of a Polygon	a) two lines b) two segments
B) Operations for computing geometric results		
7) intersection point	intersection problems	a) two segments b) two lines c) one line, one segment
8) midpoint of the circuit for three points	Voronoi diagrams	a) three exact points

by pairs, triples etc. of points. If it happens that the geometric results are representable as machine numbers in the floating–point system, these numbers constitute the desired output of a geometric algorithm. Otherwise we want to obtain a best possible approximation.

The number of geometric primitives, which suffice for the solution of two–dimensional geometric problems, is rather small. By a careful inspection of most algorithms described in [PS] we extracted eight geometric operations, which are sufficient for the numerically stable solution of problems like point–location, computation of the convex hull resp. the Voronoi–Diagram for a given point–set, kernel of a Polygon, intersection–problems.

Table 1 shows these operations and gives an overview about algorithms where they occur. Of course, the representation of the involved operands, i. e. the geometric objects, may differ. How to handle the same geometric operation numerically, crucially depends on how the respective operands are represented.

Exact points are given by their (x, y)–coordinates (as machine–numbers), lines are given by two exact points on the line, segments by their two (exact) endpoints, and intersection points are given by two lines, two segments, or a line and a segment.

3 Reporting Pairs of Intersecting Line Segments Correctly

Before considering this special problem we want to elaborate the main ideas for the implementation of geometric algorithms in general.

First of all we presuppose that all original data, i. e. the input of the geometric algorithm can be represented exactly by floating–point numbers. Problems by uncertain and only approximately representable original data shall not be considered. Our method to achieve a numerically stable implementation of a geometric algorithm can be summarized as follows:

- All *test operations* will be executed *exactly*, and

- all *real operations* to compute *geometric results* will be evaluated *lsba*, i. e. with least–significant–bit–accuracy.

The exact evaluation of a test operation and the lsba–evaluation of a real operation require to represent all geometric objects that occur as "intermediate results" in geometric operations by original data. Fortunately the computational depth of geometric algorithms is rather low as a look into the literature shows (cf. the textbook [PS], for example). In order to avoid redundancy it may be favorable to use data structures with pointers to original data (cf. the following example).

Clearly, a necessary access to original data by dereferencing a chain of pointers may slow down some geometric operations considerably. For example, a quick comparison of such objects is impossible. Because the same geometric objects may occur in many different operations it may be appropriate to choose for those geometric objects a uniform double representation: First, as the lsba–evaluated result of the operation to compute the object, and, second, as a "symbolic", exact result, i. e. as a sequence of pointers to original data. As we will see soon, this double representation of geometric objects allows also a much more efficient execution of test operations in many cases.

The exact execution of test operations can be transformed into the exact computation of the signum of one (sometimes more than one) real function. We obtain the exact signum of the function–values by a *lsba–evaluation* of the function which is defined in the following way:

Given an n–place real function $f : R^n \to R$, $(u_1,\dots,u_n) \to f(u_1,\dots,u_n)$. Let $\boxed{f}$ denote the evaluation of f on the computer. The evaluation of f is called *least–significant–bit–accurate* if:

$$\boxed{f}(x_1,\dots,x_n) = \square\, (f(x_1,\dots,x_n)), \text{ for all } x_1,\dots,x_n \in M,$$

where $\square$ is an antisymmetric rounding of the real numbers (see [KM]).

Thus the least–significant–bit–accurate evaluation (*lsba–evaluation*, for short) is the best possible representation of the exact value of f within a floating–point system. Further details of the lsba–evalution of arithmetric expression are given in [B].

The properties of an antisymmetric rounding together with the additional property $\square\, a = 0 \implies a = 0$ suffice to imply

$$sgn(f(x_1,\ldots,x_m)) = sgn(\square\,(f(x_2,\ldots,x_m))).$$

Now we want to show by an example, how we can speed up the execution of test operations by lsba–evaluated data. Let us assume that we want to determine for two intersection points $p = (p_x, p_y)$, $q = (q_x, q_y)$ in the plane their x–order. (Here, p_x, p_y, q_x, q_y denote reals, not necessarily machine numbers!) This requires to compute the exact signum of $p_x - q_x$. Observe that the coordinates of intersection points cannot be represented exactly by floating–point numbers in general. Therefore only the lsba–coordinates of $p : (\square\,(p_x), \square\,(p_y))$ resp. $q : (\square\,(q_x), \square\,(q_y))$ are available. However we obtain the following implications by the monotonicity of the rounding $\square$:

$$\square\,(p_x) < \square\,(q_x) \implies p_x < q_x$$
$$\square\,(p_x) > \square\,(q_x) \implies p_x > q_{\tilde{x}}$$

But $\square\,(p_x) = \square\,(q_x)$ does *not* imply $p_x = q_x$ in general.

Only in the latter (very improbable) case it is necessary to compute $\square\,(p_x - q_x)$ by referring to the original data which represent the two intersection points p and q. Other test operations occurring in geometric algorithms can be treated in a similar manner.

What kind of functions have to be evaluated in order to compute geometric objects? Of course the type of function crucially depends on the geometric objects. But observe, that for objects as original points, lines, segments (defined by 2 original points), intersection points of them, perpendiculars for two original or intersection points, outercircle mid–points, and many more, the usual operations to compute these objects always yield arithmetic expressions in $+, -, *, /$. Especially for test operations arithmetic expressions in $+, -, *$ are already sufficient, since an arbitrary arithmetic expression f in $+, -, *, /$ can be transformed by algebraic transformations to the quotient of two arithmetic expressions g, h in $+, -, * : f = \frac{g}{h}$. Thus we obtain $sgn(f) = sgn(g) * sgn(h)$ and the exact evaluation of $sgn(f)$ can be obtained by the lsba–evaluation of two arithmetic expressions in $+, -, *$.

Now we want to return to our sample application: Report intersecting pairs of line segments. As data–structures for the geometric objects occuring in the scan–line algorithm for solving the problem we choose:

```
point = record x, y: real end; {for original–points}
segment = record p, q: point end;
intersection–point = record x, y: real; {lsba coordinates}
                            S1, S2: ↑segment
                            {pointers to the defining segments }
                     end;
```

The *geometric objects* computed by the algorithm as output results are intersection points of pairs of line segments. Computing an intersection point includes the test whether or not an intersection between two input line segments exist. It is not difficult to see that

- the intersection test can be reduced to a point–line–location test, and
- computing the two coordinates (x, y) of an intersection point (if it exists) can be computed by a lsba–evaluation of two arithmetic expressions:

$$x = \frac{(x_2 y_1 - x_1 y_2) * (x_4 - x_3) - (x_4 y_3 - x_3 y_4) * (x_2 - x_1)}{(x_4 - x_3) * (y_1 - y_2) - (x_2 - x_1) * (y_3 - y_4)}$$

$$y = \frac{(x_2 y_1 - x_1 y_2) * (y_4 - y_3) - (x_4 y_3 - x_3 y_4) * (y_2 - y_1)}{(x_4 - x_3) * (y_1 - y_2) - (x_2 - x_1) * (y_3 - y_4)}$$

(x_1, y_1) (x_4, y_4)

(x, y)

(x_3, y_3) (x_2, y_2)

The *point–line–location* test for a given line S (represented by two original points p_1, p_2) and a given original point $p_0 = (x_0, y_0)$ is implemented as follows:

p_1
(x_1, y_1)

$p_0 = (x_0, y_0)$
S

(x_2, y_2)
p_2

If $x_1 = x_2$, the exact location of p_0 can be found by a comparison of the two machine numbers, x_0 and $x_1 (= x_2)$, otherwise we consider the equation of the line through p_1, p_2:

$$y(x) = \frac{x - x_1}{x_2 - x_1} * y_2 + \frac{x_2 - x}{x_2 - x_1} * y_1.$$

W.l.o.g. we assume $x_2 > x_1$ (otherwise exchange (x_1, y_1) and (x_2, y_2)). Then by

$$y_0 = y(x_0) \iff y_0 * (x_2 - x_1) = (x_0 - x_1) * y_2 + (x_2 - x_0) * y_1$$
$$\iff f := y_0 * x_2 - y_0 * x_1 - x_0 * y_2 + x_1 * y_2 - x_2 * y_1 + x_0 * y_1 = 0$$

we have the result, that p_0 lies on the line through p_1 and p_2 (resp. below or above), iff $f = 0$ (resp. $f < 0$ or $f > 0$) holds. Obviously, it is possible to evaluate f lsba by the optimal scalar product $scalp(u, v)$ of the vectors u, v with

$$u^T = (y_0, -y_0, -y_2, y_2, -y_1, y_1) \quad \text{and} \quad v^T = (x_2, x_1, x_0, x_1, x_2, x_0)$$

The signum of $scalp(u, v)$ gives the desired exact result (see [BRUW], [KU]).

Finally we show how to determine exactly the x–order between intersection and original points. The problem is trivial for two original points of course. If we have one intersection point $S = (x, y)$ (the lsba–coordinates ($\Box$ (x), $\Box$ (y)) have been computed already) and one original point $p = (x_0, y_0)$, the problem can be solved exactly by a computation of $\Box$ $(x_0 - \Box$ $(x))$, if x_0 is different from $\Box$ (x). Thus there only remains the problem to determine the signum of $(x_0 - x)$ where $x_0 = \Box$ x.

Denoting the denominator of x by D and the enumerator of x by E we obtain:

$$sgn(x_0 - x) = sgn\left(x_0 - \frac{E}{D}\right) = sgn\left(\frac{(D * x_0 - E)}{D}\right) = sgn(D * x_0 - E) * sgn(D)$$

As $D * x_0 - E$ is an arithmetic expression in $+, -, *$, more exactly a sum of products with 3 factors, the signum can be evaluated exactly by optimal scalar product techniques [B]. D can be written as a scalar–product, thus the x–order can be determined exactly. If both points are intersection–points, the problem isn't more difficult, but, of course, the arithmetic expression becomes more complex; it's a sum of products with 5 factors, which must be evaluated.

These operations, implemented as required in the beginning of this section, suffice for a numerically stable implementation of the scan–line algorithm. In the next section we show by some experimental results, that the resulting algorithm is still reasonably efficient.

4 Test Results

Our method of implementing the geometric primitives of geometric algorithms is to reduce geometric operations to the lsba–evaluation of arithmetic expressions. This method can be assessed from two different points of view:

We can first compare the computed result with the result of a floating–point approximation. Secondly, we can ask for the efficiency measured by the run–time to carry out a lsba–evaluation of the respective arithmetic expressions on different computers for representative sets of data.

What concerns the first criterion (*quality of the result*) it is obvious that the number of significant digits for the mantissa representation of the original data plays an important role. If we can, for example, represent all input data with only 3 significant decimal digits and the computer allows floating–point arithmetic with considerably more decimal digits, a floating–point approximation may be good enough in almost all cases.

We implemented a number of geometric primitives, in particular the intersection of two line segments defined by two pairs of original points on a 68000–based microcomputer (KWS SAM 68K) which has a floating–point arithmetic with 13 significant decimal digits and a software–supported optimal scalar–product. As expected the quality of the result of the simple floating–point computation not only depends on the number of significant digits of the original data but also on the angle under which the two lines intersect: The number of wrong digits compared to the lsba–evaluated "best possible" result drastically increases with diminishing intersection angle. For more details cf. [Th].

What concerns the *efficiency* of lsba–evaluations note that a comparison of the run–time to compute a "best possible" (i. e. a correct) result with the runtime to compute an approximation (which may be completely wrong) is unfair. Nevertheless we carried out such a comparison by computing 1000 intersection points of line segments where the line segments were given by pairs of original points with a varying number of significant digits, intersection angles, and different range in the floating–point system. The following table summarizes the results of our experiment:

number of significant digits of the coordinates of the points defining the lines	average run–time (in milli–seconds) to compute an intersection of two lines on KWS SAM 68K by:	
	floating–point arithmetic with 13 significant digits	lsba– evaluation
3	17	30
6 – 9	19	36
10 – 13	21	40 – 60

The runtime of the usual floating-point approximation is almost independent of the length of the data. The lsba–evaluation, however, heavily depends on it. But we believe that the observed loss of efficiency by a factor between 2 and 3 is not a too big prize to pay for a correct implementation.

Although many data were tested for different operations, some of them much more complex than the computation of an intersection point, there never occured an over– or underflow, even for rather "unrealistic" data. This is not surprising, because usually geometric data lie in the "middle" of the floating–point system. Thus, an exponent range from −99 to 99 or something more should be sufficient for completely avoiding over– or underflow in most practical applications.

5 Conclusion

In this paper we have shown that the lsba–evaluation of arithmetic expressions provides a solid basis for a clean implementation of geometric primitives which occur in many geometric algorithms. It turns out that it is necessary to have access not only to the computed results but also to the original data from which these results are computed. This implies that in general geometric objects are associated with references to original data.

We extracted the geometric primitives from a large class of algorithms for solving geometric problems in two dimensions. These primitives were implemented such that test operations always yield the correct result and operations to compute geometric objects yield a "best possible" machine representation. A comparison of the efficiency with standard floating point arithmetic shows a loss of efficiency by a factor between two and three on a computer with a software–supported optimal scalar–product. We conjecture that the loss of efficiency for an analogous implementation of the geometric primitives needed to solve problems in three (or more) dimensions is not much larger.

From a practical point of view it is even more important to solve the numerical problems occurring in algorithms for 3–dimensional problems, as, for example, computing the intersection of solids in 3–dimensional space, rotating or translating of spatial geometric objects etc. We are quite optimistic that the method described in this paper can be considerably extended and leads to a general, solid, and mathematically clean foundation of the numerical aspects in geometric algorithms.

References

[A]   V. Akman: Unobstructed shortest paths in polyhedral environments. Lecture Notes in Computer Science, vol.251,1987

[B]   H. Böhm: Berechnung von Polynomnullstellen und Auswertung arithmetischer Ausdrücke mit garantierter maximaler Genauigkeit. PhD Thesis, Karlsruhe, December 1983.

[Ber]   G.B. Beretta: An implementation of a plane sweep–algorithm on a personal computer. Dr.–Dissertation, Eidgenössische Technische Hochschule Zürich, 1984.

[BO]   J.L. Bentley, Th. Ottmann: Algorithms for reporting and counting geometric intersecitons. IEEE Transactions on Computers, 28, p. 643–647, 1979.

[BRUW]   G. Bohlender, L.B. Rall, Ch. Ullrich, J. Wolff von Gudenberg: PASCAL–SC: A Computer Language for Scientific Computation, Academic Press (Perspectives in Computing, vol. 17), Orlando, 1987.

[F]   A.R. Forrest: Computational Geometry and Software Engineering: Towards a Geometric Computing Environment. In "State–of–the–Art in Computer Graphics", Eds. R.A. Earnshaw and D.F. Rogers, Springer, March 1987.

[GS]   L. Guibas, I. Stolfi: Primitives for the Manipulation of General Subdivisions and the Computation of Voronoi–Diagrams. ACM Transactions on Graphics, 4 (2), 74–123, 1985.

[KM]   U. Kulisch, W.L. Miranker: Computer Arithmetic in Theory and Practice. Academic Press, New York, 1981.

[KU]   U. Kulisch, Ch. Ullrich (Eds.): Wissenschaftliches Rechnen und Programmiersprachen. Berichte des German Chapter of the ACM, vol. 10, Teubner, Stuttgart, 1982.

[M]   Mathlab Group: MACSYMA Reference Manual, 2 vols., Lab. for Computer Science, Massachusetts Inst. of Technology, Cambridge, MA, 1983.

[OTU]   Th. Ottmann, G. Thiemt, Ch. Ullrich: Numerical Stability of simple Geometric Algorithms in the plane. Lecture Notes in Computer Science, Vol. 270: Computation Theory and Logic, 277–293, Springer, Berlin, 1987.

[PS]   F.P. Preparata, M.I. Shamos: Computational Geometry. Springer, N.Y., 1985.

[Ra]   L. Ramshaw: CSL Notebook Entry: The Braiding of Floating Point Lines. Unpublished note, Xerox PARC, Oct. 1982.

[Th]   G. Thiemt: Die numerische Stabilität geometrischer Algorithmen. Diploma Thesis, Karlsruhe, Oktober 1986.

Prof. Dr. Th. Ottmann   Dipl. math. G. Thiemt   Prof. Dr. Ch. Ullrich
Institut für Informatik   Computer Vision   Institut für Informatik
Universität Freiburg   Deutschland GmbH   Universität Basel
Rheinstrasse 10–12   Riedstrasse 25   Mittlere Strasse 142
D–7800 Freiburg   D–7302 Ostfildern 1   CH–4056 Basel
Federal Republic   Federal Republic   Switzerland
of Germany   of Germany

III. Improving the Tools

Computing. Suppl. 6, 139–148 (1988)

Precise Evaluation of Polynomials in Several Variables

R. Lohner, Karlsruhe

Abstract — Zusammenfassung

Precise Evaluation of Polynomials in Several Variables. In this paper an algorithm is presented which computes the values of polynomials in several variables with high accuracy when the computation is done in a floating-point system using a precise scalar product and directed rounding. Furthermore the value of the polynomial is enclosed in narrow bounds. Numerical examples demonstrate the high precision of the results and show that traditional algorithms (e.g. nested Horner's schemes) can fail completely for this problem. The algorithm presented here is a generalization of Böhm's algorithm for one-dimensional polynomials, [4], [5].

AMS Subject Classification: 65 G 10.

Key words: Polynomial in several variables, exact scalar product, interval arithmetic, staggered correction format.

Genaue Auswertung von Polynomen in mehreren Variablen. In dieser Arbeit wird ein Algorithmus vorgestellt, mit dem der Wert von Polynomen in mehreren Variablen mit hoher Genauigkeit berechnet werden kann, wobei die Rechnung in einem Gleitpunktsystem mit genauem Skalarprodukt und gerichteter Rundung durchgeführt wird. Überdies wird der Polynomwert in engen Schranken eingeschlossen. Numerische Beispiele demonstrieren die hohe Genauigkeit der Ergebnisse und zeigen, daß traditionelle Verfahren (z.B. geschachteltes Hornerschema) bei diesem Problem völlig versagen können. Der hier vorgestellte Algorithmus ist eine Verallgemeinerung des Algorithmus von Böhm für eindimensionale Polynome, [4], [5].

1. Introduction

The evaluation of polynomials in a floating-point system on a computer is a nontrivial task even in the one-dimensional case as can be seen by many examples, [4], [5], [12]. In [4] H. Böhm has presented an algorithm which computes bounds of maximum accuracy for the value $p(x)$ of a one-dimensional polynomial with floating-point coefficients and for a floating-point argument x. In this paper we will generalize this algorithm in two steps: *first* the coefficients of the polynomial need *no longer* be *floating-point numbers* and *second* we use the thus modified algorithm for evaluating *polynomials in several variables*.

We use a floating-point system **S** with base B and mantissa length t. The set of all intervals with bounds in **S** is denoted by **IS**. The symbols $\square$, $\triangle$, ∇ resp. $\diamondsuit$ denote a monotone antisymmetric, upwardly directed, downwardly directed resp. interval rounding. The corresponding rounded floating-point operations are denoted analogously, e.g. $\boxplus$, $\triangle$, ∇, $\diamondsuit$, etc.

Furthermore we assume that all floating-point operations are of maximum accuracy and that we have available a precise scalar-product for vectors of floating-point numbers or, equivalently, that we have a long accumulator, i.e. for $x = (x_i)$, $y = (y_i)$, x_i, $y_i \in S$, $i = 1, ..., n$, we assume that we can compute $x \boxdot y$

$$:= \square \left(\sum_{i=1}^{n} x_i \cdot y_i \right) \text{ with one rounding only (and similarly for the other roundings).}$$

All these requirements are fulfilled e.g. by the programming languages PASCAL-SC (which was used to compute the numerical examples in Section 4) as well as FORTRAN-SC and by the ACRITH subroutine library on IBM/370 computers. For more details concerning this kind of arithmetic see [1], [3]−[7], [9]−[12].

We will also use the notation $\boxed{m}(X)$ for the rounded midpoint of an interval $X \in \mathbf{IS}$, i.e. if $X = [\underline{X}, \bar{X}]$ then $\boxed{m}(X) = \square((\underline{X} + \bar{X})/2)$, $\operatorname{diam}(X) = \bar{X} - \underline{X}$ is the diameter of the interval X and with $\min|X| = \min\{|x| \,|\, x \in X\}$ we denote the smallest absolute value of all elements in X.

The enclosure of a real number $a \in \mathbb{R}$ not just by an interval of $\mathbf{IS}$, but by a sum of r floating-point numbers $a_s \in S$ plus an interval $A \in \mathbf{IS}$

$$a \in \sum_{s=1}^{r} a_s + A, \qquad a_s \in S, \qquad A \in \mathbf{IS}, \tag{1}$$

will be called an enclosure is *staggered correction format*. This format was introduced in [2] and [13] and is also studied in detail in [8]. We note that this format requires to store all floating-point numbers a_s and the interval A in the computer. The sum in (1) is not formed explicitly in the floating-point system since otherwise the gain in accuracy (up to $r + 1$ mantissas) would be reduced again to one mantissa only. Thus a staggered correction format is essentially a simulation of a dynamic multiple precision data type. Whenever arithmetic operations have to be performed with an enclosure of the form (1), these operations must use the precise scalar-product or the long accumulator, see e.g. Algorithm 2 in Section 2 (we do not need multiplication or division for staggered correction data).

Finally we note that if $x \in X$ then for any approximation $\tilde{x} \in X$ of x the inequality

$$\operatorname{diam}(X) \le \varepsilon \cdot \min|X|, \qquad \text{for } 0 \notin X, \qquad \varepsilon \ge 0, \tag{2}$$

obviously implies that the relative error of $\tilde{x}$ is less than or equal to ε. This holds regardless of X being a real interval, an interval in $\mathbf{IS}$ or in staggered correction format.

After a short review of Böhm's algorithm for one-dimensional polynomials we present in Section 2 a modification for polynomials whose coefficients are intervals in staggered correction format. This modified algorithm then will be used in Section 3 for the evaluation of polynomials in several variables.

2. Polynomials in One Variable

The algorithm for evaluating a one-dimensional polynomial with maximum accuracy ([4]) is based on reformulating Horner's scheme as a system of linear equations: let

$$p(x) = \sum_{i=0}^{n} a_i x^i = a_0 + a_1 x + \dots + a_n x^n$$

$$= a_0 + x(a_1 + x(\dots + x(a_{n-1} + x a_n)\dots)),$$

then we introduce for each intermediate result a new auxiliary variable x_i (starting from $i = n$):

$$x_0 = a_0 + x \cdot x_1,$$
$$\vdots$$
$$x_{n-1} = a_{n-1} + x \cdot x_n,$$
$$x_n = a_n.$$

This is a system of linear equations for the auxiliary variables and we have $x_0 = p(x)$. In matrix-vector notation the system reads:

$$
\underbrace{\begin{pmatrix} 1 & -x & & & \\ & 1 & -x & & 0 \\ & & \ddots & \ddots & \\ & 0 & & 1 & -x \\ & & & & 1 \end{pmatrix}}_{A} \cdot \underbrace{\begin{pmatrix} x_0 \\ x_1 \\ \vdots \\ x_{n-1} \\ x_n \end{pmatrix}}_{y} = \underbrace{\begin{pmatrix} a_0 \\ a_1 \\ \vdots \\ a_{n-1} \\ a_n \end{pmatrix}}_{b}. \tag{3}
$$

For systems of linear equations the residual iteration technique works excellent in floating-point arithmetic if the precise scalar-product or the long accumulator is used to compute critical terms such as the residue $b - A\tilde{y}$ for an approximate solution $\tilde{y}$ (see [4], [5], [11], [12] for details). It is this residual iteration for equation (3) which is realized in the following algorithm of Böhm ([4], [5]).

In step 1) of Algorithm 1 a floating-point approximation $x_i^{(0)}$ for each of the auxiliary variables x_i is computed by the usual floating-point Horner's scheme.

In step 2) then the residual iteration for system (3) is performed where enclosures of the x_i are computed in staggered correction form: in the k-th iteration x_i is approximated by the sum of the floating-point numbers $x_i^{(j)}$, $j = 0, \dots, k-1$, the corresponding residue of System (3) is enclosed in the intervals $D_i^{(k)}$ with maximum accuracy and the error of this approximation is enclosed in $X_i^{(k)}$ obtained by solving the residual equation using interval arithmetic. The iteration is stopped when the desired accuracy is reached or when after a minimum of k_{max} ($= 10$, say) iterations underflow occurs (The termination criterion in [4] is slightly different, but essentially equivalent). In any case the resulting interval $Y^{(k)}$ is an enclosure of $p(x)$ (although possibly not of maximal accuracy if an underflow occurred).

Algorithm 1.

(Evaluation of a one-dimensional polynomial with floating-point coefficients)

$$p(x) = \sum_{i=0}^{n} a_i x^i, \quad a_i, x \in S.$$

1) Approximation (Horner's scheme)

$x_n^{(0)} := a_n$
for $i := n-1$ **downto** 0 **do** $\quad x_i^{(0)} := a_i \boxplus x \boxdot x_{i+1}^{(0)}$
$X^{(0)} := x^{(0)}$

2) Iteration

$k := 0$
repeat
$\quad k := k+1$ *(Midpoint)*
$\quad x^{(k-1)} := \boxed{m}(X^{(k-1)})$
$\quad D_n^{(k)} := 0$
$\quad$ **for** $i := n-1$ **downto** 0 **do**

$$D_i^{(k)} := \diamondsuit\left(a_i - \sum_{j=0}^{k-1} x_i^{(j)} + x \cdot \sum_{j=0}^{k-1} x_{i+1}^{(j)}\right) \quad \textit{(Enclosure of residue)}$$

$\quad X_n^{(k)} := D_n^{(k)}$
$\quad$ **for** $i := n-1$ **downto** 0 **do** $\qquad$ *(Solve $AX = D$ with*
$\qquad X_i^{(k)} := D_i^{(k)} \diamondsuit x \diamondsuit X_{i+1}^{(k)}$ $\qquad$ *interval arithmetic)*

$$Y^{(k)} := \diamondsuit\left(\sum_{j=0}^{k-1} x_0^{(j)} + X_0^{(k)}\right)$$

until $\operatorname{diam}(X_0^{(k)}) \leq \varepsilon \cdot \min |Y^{(k)}|$ **or** (underflow **and** $k \geq k_{\max}$)

3) Results

It is proved that $p(x) \in Y^{(k)}$ and the relative error of the result is less than ε provided $0 \notin Y^{(k)}$ and no underflow occurred (and no overflow).

With only four minor modifications it is possible to apply this algorithm to polynomials whose coefficients are given in staggered correction format and to compute the value of the polynomial also in staggered correction format. We list the modified algorithm as Algorithm 2.

The modifications are:

(i) in step 1) the approximations $x_i^{(0)}$ are computed using only the values $a_i^{(1)}$;
(ii) in the computation of the enclosures $D_i^{(k)}$ the staggered correction format of the coefficients has to be used instead of only floating-point numbers as in Algorithm 1 (therefore $D_n^{(k)} \neq 0$ in general);
(iii) the computation of $Y^{(k)}$ is not done explicitly since the sum of the numbers $x_0^{(j)}$ plus the interval $X_0^{(k)}$ is already the desired staggered correction format of the result $p(x)$;

(iv) the termination criterion must be changed since now it is possible that no underflow occurs and the desired accuracy cannot be reached when the diameters of the coefficients $\operatorname{diam}(a_i)=\operatorname{diam}(A_i)$ are too large.

Remark. As pointed out in modification (iv) it is possible that the accuracy cannot be increased even if underflow does not yet occur. If, however, $\operatorname{diam}(A_i)=0$ for all $i=0,\dots,n$, then the accuracy can be increased as in Algorithm 1 until underflow occurs.

Algorithm 2.

(Evaluation of a one-dimensional polynomial with coefficients in staggered correction format)

$$p(x)=\sum_{i=0}^{n} a_i\, x^i,\quad a_i=\sum_{s=1}^{r} a_i^{(s)}+A_i,\quad a_i^{(s)}, x\in S,\quad A_i\in IS.$$

1) Approximation (Horner's scheme)

$x_n^{(0)}:=a_n^{(1)}$
for $i:=n-1$ **downto** 0 **do** $\quad x_i^{(0)}:=a_i^{\{1} \boxplus x \boxdot x_{i+1}^{(0)}$
$X^{(0)}:=x^{(0)}$

2) Iteration

$k:=0$
repeat
$\qquad k:=k+1$
$\qquad x^{(k-1)}:=\boxed{\text{m}}(X^{(k-1)})$ $\hfill$ (*Midpoint*)
$$D_n^{(k)}:=\Diamond\left(\sum_{s=1}^{r} a_n^{(s)}+A_n-\sum_{j=0}^{k-1} x_n^{(j)}\right)$$
$\qquad$ **for** $i:=n-1$ **downto** 0 **do**
$$D_i^{(k)}:=\Diamond\left(\sum_{s=1}^{r} a_i^{(s)}+A_i-\sum_{j=0}^{k-1} x_i^{(j)}+x\cdot\sum_{j=0}^{k-1} x_{i+1}^{(j)}\right)$$
$\hfill$ (*Enclosure of residue*)
$\qquad X_n^{(k)}:=D_n^{(k)}$
$\qquad$ **for** $i:=n-1$ **downto** 0 **do** $\hfill$ (*Solve* $AX=D$ *with*
$\qquad\qquad X_i^{(k)}:=D_i^{(k)}\Diamondplus x \Diamond X_{i+1}^{(k)}$ $\hfill$ *interval arithmetic*)
$\qquad Y^{(k)}:=\sum_{j=0}^{k-1} x_0^{(j)}+X_0^{(k)}$
until $\operatorname{diam}(X_0^{(k)})\le\varepsilon\cdot\min|Y^{(k)}|$ **or** $(k\ge k_{\max})$

3) Result

It is proved that $p(x)\in\sum_{j=0}^{k-1} x_0^{(j)}+X_0^{(k)}$ and the relative error of the result is less than ε provided $0\notin Y^{(k)}$ and $k<k_{\max}$ (and no overflow occurred).

3. Polynomials in Several Variables

To avoid too many indices we present the evaluation of polynomials in several variables only in the case of $N=2$ and $N=3$ variables which we denote by x, y and z. The general case is then completely obvious and involves only more typing effort. Also we use polynomials with maximal degree in each variable since then in the nested summations the summation indices are independent of each other. The method works in the same way for polynomials with maximal total degree, but then the indices of the inner sums depend on those of the outer sums, i.e. we have a rectangular coefficient-array in the first case and a triangular coefficient-array in the second case.

The value of the two-dimensional polynomial of degree n in x and degree m in y

$$p(x, y) = \sum_{j=0}^{m} \sum_{i=0}^{n} a_{ij} x^i y^j = \sum_{j=0}^{m} \left(\sum_{i=0}^{n} a_{ij} x^i \right) y^j \tag{4}$$

can be obtained by successively computing the values of the $m+2$ one-dimensional polynomials:

$$b_j := b_j(x) := \sum_{i=0}^{n} a_{ij} x^i, \quad j = 0, \dots, m,$$

$$p(x, y) = \sum_{j=0}^{m} b_j y^j \tag{5}$$

or, reversing the order of summation in (4), by the $n+2$ one-dimensional polynomials

$$c_i := c_i(y) := \sum_{j=0}^{m} a_{ij} y^j, \quad i = 0, \dots, n,$$

$$p(x, y) = \sum_{i=0}^{n} c_i x^i. \tag{6}$$

Similarly for three-dimensional polynomials

$$p(x, y, z) = \sum_{k=0}^{l} \sum_{j=0}^{m} \sum_{i=0}^{n} a_{ijk} x^i y^j z^k = \sum_{k=0}^{l} \left(\sum_{j=0}^{m} \left(\sum_{i=0}^{n} a_{ijk} x^i \right) y^j \right) z^k \tag{7}$$

$p(x, y, z)$ can be obtained by the evaluation of the $(m+2)(l+1)$ nested one-dimensional polynomials

$$b_{jk} := b_{jk}(x) := \sum_{i=0}^{n} a_{ijk} x^i, \quad j = 0, \dots, m, \quad k = 0, \dots, l,$$

$$c_k := c_k(x, y) := \sum_{j=0}^{m} b_{jk} y^j, \quad k = 0, \dots, l, \tag{8}$$

$$p(x, y, z) = \sum_{k=0}^{l} c_k z^k,$$

where also several possibilities exist depending on the order of summation in (7).

Now it is already obvious how we can use the result of Section 2. For the evaluation of a polynomial in several variables we first have to chose an order of summation and then evaluate the sums from inside to outside. However, since the values of the "inner" polynomials will become the coefficients of the next "outer" polynomial it generally would be a great loss of accuracy if only Algorithm 1 would be used, since then already the second inner polynomial would have coefficients in **IS**.

Algorithm 2, however, produces its output in staggered correction format and thus the increase of accuracy of the coefficients of the first inner polynomial is limited only by underflow (if the coefficients of p are in **S**).

Thus, if we want to compute the value of a three-dimensional polynomial, say, and require a relative error ε_3 of the result then we use the following algorithm where δ is a number less than 1.

Algorithm 3.

(Evaluation of a three-dimensional polynomial)

$$p(x, y, z) = \sum_{k=0}^{l} \sum_{j=0}^{m} \sum_{i=0}^{n} a_{ijk} x^i y^j z^k = \sum_{k=0}^{l} \left(\sum_{j=0}^{m} \left(\sum_{i=0}^{n} a_{ijk} x^i \right) y^j \right) z^k$$

1) Compute the values of the polynomials

$$b_{jk} := b_{jk}(x) := \sum_{i=0}^{n} a_{ijk} x^i, \quad j=0, \ldots, m, \quad k=0, \ldots, l,$$

using Algorithm 2 with $\varepsilon = \varepsilon_1 := \delta^2 \varepsilon_3$. The results b_{jk} are in staggered correction format.

2) Compute the values of the polynomials

$$c_k := c_k(x, y) := \sum_{j=0}^{m} b_{jk} y^j, \quad k=0, \ldots, l,$$

using Algorithm 2 with b_{jk} in staggered correction format as obtained from step 1) and take $\varepsilon = \varepsilon_2 := \delta \varepsilon_3$. The results c_k are in staggered correction format.

3) Compute the value of the polynomial

$$p(x, y, z) = \sum_{k=0}^{l} c_k z^k,$$

using Algorithm 2 with c_k in staggered correction format as obtained from step 2) and take $\varepsilon = \varepsilon_3$. The result $p(x, y, z)$ is in staggered correction format and may be rounded into **IS** if desired.

4) If the result is of desired accuracy then stop, otherwise reduce δ, e.g. set $\delta := \delta_0\,\delta$ for a fixed $\delta_0 < 1$ or set $\delta := \delta^2$ and start again with step 1).

In practice it turned out that $\delta \approx B^{-t/2}$ was an acceptable value such that in most cases restarting Algorithm 3 by step 4) was not necessary.

Finally, we remark that this algorithm is expected to be economical only in the case if the coefficient-array of the polynomial is dense, i.e. if only few coefficients are equal to zero. In the other case, if only a few coefficients are different from zero, then probably more general formula evaluation techniques such as described in [4], [5] and [12] are more economical.

4. Numerical Examples

In this section we present some numerical results obtained with Algorithm 3. We compare these results with the approximations which were computed by substituting a floating-point Horner's scheme for Algorithm 2 in the polynomial evaluations of Algorithm 3.

All computations were done in PASCAL-SC on a ATARI 1040STF using a 13 digit decimal floating-point arithmetic with precise scalar product (for details see [10]). Intervals will be written in an obvious short notation in the sequel.

Example 1.

$$p(x, y) = 9x^4 - y^4 + 2y^2$$

The exact value for $x = 10864$ and $y = 18817$ is $p(x, y) = 1$ which is also obtained exactly by Algorithm 3. With Horner's scheme we obtain the value $1.0\mathrm{E}+05$.

Example 2.

$$p(x, y) = 83521\,y^8 + 578\,x^2\,y^4 - 2x^4 + 2x^6 - x^8$$

The exact value for $x = 9478657$ and $y = 2298912$ is $p(x, y) = -179689877047297$. With Algorithm 3 we obtained the enclosure $p(x, y) \in -1.79689877047\frac{2}{3}\mathrm{E}+14$. The floating-point Horner's scheme, however, yields the approximation $-2.0\mathrm{E}+43$.

Example 3.

$$p(x, y) = -x^6 - 3x^5 y + 5x^3 y^3 - 3x y^5 + y^6$$

The exact value of the polynomial for $x = a_i$ and $y = a_{i+1}$ is $p(x, y) = (-1)^i$ where a_i and a_{i+1} are two sequential *Fibonacci numbers*:

$$a_0 = 0, \quad a_1 = 1, \quad a_{i+2} = a_{i+1} + a_i, \quad i \geq 0. \tag{9}$$

In the following Table 1 we list some results for this polynomial.

Table 1

i	a_i	a_{i+1}	Horner	Algorithm 3
10	55	89	1	1
13	233	377	0	-1
16	987	1597	$4\,\mathrm{E}+05$	1
19	4181	6765	0	-1
22	17711	28657	$2.5\,\mathrm{E}+14$	1
45	1134903170	1836311903	$-6\,\mathrm{E}+42$	-1
62	4052739537881	6557470319842	$7\,\mathrm{E}+64$	1

Table 2

i	a_{i-1}	a_i	a_{i+1}	Horner	Algorithm 3
11	55	89	144	-1	-1
14	233	377	610	$2\,\mathrm{E}+04$	1
17	987	1597	2584	$2\,\mathrm{E}+08$	-1
20	4181	6765	10946	0	1
23	17711	28657	46368	$-1\,\mathrm{E}+15$	-1
46	1134903170	1836311903	2971215073	$1\,\mathrm{E}+44$	1
62	2504730781961	4052739537881	6557470319842	$9\,\mathrm{E}+64$	1

Example 4.

$$p(x, y, z) = (x^2 - 1) \cdot (y^2 - 4) \cdot (z^2 - 9)$$
$$= x^2 y^2 z^2 - 4 x^2 z^2 - y^2 z^2 + 4 z^2 - 9 x^2 y^2 + 36 x^2 + 9 y^2 - 36$$

Obviously we have $p(\pm 1, y, z) = p(x, \pm 2, z) = p(x, y, \pm 3) = 0$ for all x, y, $z \in \mathbb{R}$. However, with floating-point Horner's scheme we obtain the approximations

$$p(11111, 22222, 3) \approx 2.0\,\mathrm{E}+05,$$
$$p(987654, 123456, 3) \approx 1.0\,\mathrm{E}+11,$$
$$p(123.456789, 123.456789, 3) \approx 1.0\,\mathrm{E}-03,$$

whereas Algorithm 3 in all cases yields the correct results 0.

Example 5.

$$p(x, y, z) = 6 x y^5 + 15 x^2 y^4 + 20 x^3 y^3 + 15 x^4 y^2$$
$$+ 6 x^5 y + x^6 - 3 y z^5 + 5 y^3 z^3 - 3 y^5 z$$

This polynomial is obtained from the one in Example 3 using the definition of the Fibonacci sequence to obtain one more variable. Here too, the exact value of the polynomial for $x = a_{i-1}$, $y = a_i$ and $z = a_{i+1}$ is $p(x, y, z) = (-1)^i$ where again the a_i are the Fibonacci numbers (9). Results for this polynomial are listed in Table 2.

References

[1] Alefeld, G., Herzberger, J.: Introduction to Interval Computations. New York: Academic Press 1983.

[2] Auzinger, W., Stetter, H. J.: Accurate Arithmetic Results for Decimal Data and Non-Decimal Computers, Computing 35, 1985.

[3] Bleher, J. H., Kulisch, U., Metzger, M., Rump, S. M., Ullrich, Ch., Walter, W.: FORTRAN-SC: A Study of a FORTRAN Extension of Engineering/Scientific Computation with Access to ACRITH. Computing 39, 93–110 (1987).

[4] Böhm, H.: Berechnung von Polynomnullstellen und Auswertung arithmetischer Ausdrücke mit garantierter maximaler Genauigkeit. Dissertation, Universität Karlsruhe: 1983.

[5] Böhm, H.: Evaluation of Arithmetic Expressions with Maximum Accuracy. p. 121 in [6].

[6] Grüner, K.: Solving Complex Problems for Polynomials and Linear Systems with Verified High Accuracy. p. 199 in [5].

[7] Kaucher, E., Kulisch, U., Ullrich, Ch. (eds.): Computerarithmetic, Scientific Computation and Programming Languages. Stuttgart: B. G. Teubner 1987.

[8] Klotz, G.: Faktorisierung von Matrizen mit maximaler Genauigkeit. Dissertation, Universität Karlsruhe: 1987.

[9] Kulisch, U. (ed.): PASCAL-SC, A Pascal Extension for Scientific Computation, Information Manual and Floppy Disks. Version IBM PC/AT; operating system DOS. Stuttgart: B. G. Teubner (Wiley-Teubner series in Computer Science) 1987.

[10] Kulisch, U. (ed.): PASCAL-SC, A Pascal Extension for Scientific Computation, Information Manual and Floppy Disks. Version ATARI ST. Stuttgart: B. G. Teubner 1987.

[11] Kulisch, U., Miranker, W. (eds.): A New Approach to Scientific Computation. New York: Academic Press 1983.

[12] Rump, S.: Solving Algebraic Problems with High Accuracy. p. 53 in [6].

[13] Stetter, H. J.: Sequential Defect Correction for High Accuracy Floating-Point Algorithms. Lecture Notes in Mathematics, Vol. 1066, p 186–202, 1984.

Rudolf Lohner
Institut für Angewandte Mathematik
Universität Karlsruhe
Kaiserstrasse 12
D-7500 Karlsruhe 1
Federal Republic of Germany

Computing. Suppl. 6, 149–158 (1988)

Evaluation of Arithmetic Expressions
with Guaranteed High Accuracy [1]

H. C. Fischer, G. Schumacher, Karlsruhe, and **R. Haggenmüller,** München

Abstract — Zusammenfassung

Evaluation of Arithmetic Expressions with Guaranteed High Accuracy. The advantage of having available an exact scalar product in numerical computations has already been mentioned by many authors. In this paper we use this tool to develop an algorithm for the evaluation of rational expressions in several variables with high guaranteed accuracy. An extension for complex expressions, vector-matrix expressions and expressions with standard functions and interval arguments is also given.

AMS subject classification: 65G10

Keywords: formula evaluation, interval arguments, matrix expressions, verified results, high accuracy

Auswertung arithmetischer Ausdrücke mit garantierter, hoher Genauigkeit. Die Vorteile eines exakten Skalarprodukts für numerische Berechnungen sind bereits von vielen Autoren erwähnt worden. Mit seiner Hilfe wird in dieser Arbeit ein Algorithmus für die Auswertung von rationalen Ausdrücken in einer oder mehreren Variablen mit garantierter, hoher Genauigkeit entwickelt. Eine Erweiterung für komplexe Ausdrücke, Vektor bzw. Matrixausdrücke sowie Ausdrücke mit Standardfunktionen und Intervallargumenten wird ebenfalls gegeben.

0. Introduction

The requirement of having mathematical models which approximate the real world in a very smooth way has lead to more and more complicated formulae. Many of those formulae used in modern engineering are not created by hand but by symbolic manipulations on a computer. However, the numerical calculations following the symbolic manipulations are usually done using floating-point arithmetic. An error analysis for these computations should be done by the computer automatically.

In this paper we consider a partial problem of this context, the evaluation of arithmetic expressions. Especially for algorithms using the techniques of iterative refinement, the accurate evaluation of expressions is important, since the computed results of residue terms may be inaccurate owing to cancellations.

[1] This work was supported by the Commission of the European Communities, ESPRIT Project No. 1072 (DIAMOND) and the Ministerium für Wirtschaft, Mittelstand und Technologie, Baden-Württemberg.

The first ad hoc idea to deal with this problem is to use multiple-precision calculation. Unfortunatley, the hardware of nearly all computing machines does not support such a calculation and so it has to be simulated. As part of the ESPRIT project DIAMOND we examined the possibility of using an exact scalar product as introduced in [8] to simulate a multiple-precision calculation. The procedure developed in the following is based on this idea. It is not done by providing multiple-precision operators, such as $+$, $-$, $*$, $/$, but by embedding the multiple-precision calculation into an iterative correction process. This is equivalent to a recomputation of the formula with increasing precision, but it allows to reduce the over-all computing time by using more store.

We only mention that in the FORTRAN-libraries ACRITH (IBM) [10] and ARITHMOS (Siemens) [11] routines are contained providing guaranteed bounds for the value of an arithmetic expression defined by $+$, $-$, $*$, $/$ and powers with constant integer exponents.

1. A General Strategy for the Evaluation of Expressions

We first consider the evaluation of a function $f: D_1 \times \ldots \times D_m \to \mathbb{R}$, $D_i \subseteq \mathbb{R}$ $(i = 1, \ldots, m)$, which is represented by an arithmetic expression in m variables $x_1, \ldots, x_m$. The function value $f(x_1, \ldots, x_m)$ can be computed in $\mathbb{R}$ by a successive calculation process "from left to right", performing one operation after the other and using intermediate results z_k. Let n be the number of operations in f. So we have

$$
\begin{aligned}
z_1 &:= f_1(x_1, \ldots, x_m) \\
z_2 &:= f_2(x_1, \ldots, x_m, z_1) \\
&\ \vdots \\
f = z_n &:= f_n(x_1, \ldots, x_m, z_1, \ldots, z_{n-1}),
\end{aligned}
\tag{1}
$$

where f_k $(k = 1, \ldots, n)$ is a function representing an operation for only one or two of its arguments.

Example:

$$
\begin{aligned}
f(x_1, x_2, x_3, x_4, x_5) &= (x_1 + x_2)^3 \cdot (x_3 - x_4)/x_5 \\
z_1 &:= x_1 + x_2 ; \\
z_2 &:= z_1^3 ; \\
z_3 &:= x_3 - x_4 ; \\
z_4 &:= z_2 * z_3 ; \\
z_5 &:= z_4/x_5 ;
\end{aligned}
$$

The basic idea is now to iterate the evaluation of (1) with increasing precision until we have a sufficient good approximation of f. The quality of the approximation is controlled by computation of inclusions for the errors of the intermediate

and final results. We also want to use the exact scalar product to compute
the residue terms

$$z_k - f_k(x_1, \ldots, x_m, z_1, \ldots, z_{k-1}).$$

Unfortunately, this is difficult for division and some other possible f_k's and
so we will use an implicit form of (1):

$$
\begin{aligned}
g_1(z_1) \quad &= 0 \\
g_2(z_1, z_2) \quad &= 0 \\
&\;\;\vdots \\
g_n(z_1, \ldots, z_n) &= 0,
\end{aligned}
\tag{2}
$$

where $g_k : D_1 \times \ldots \times D_k \to \mathbb{R}$, $D_i \subset \mathbb{R}$, $k=1, \ldots, n$, $i=1, \ldots, k$. For abbreviation,
we also write $D := D_1 \times \ldots \times D_n$. We have ommitted the variables $x_1, \ldots, x_m$
which are common arguments of all g_k and are considered as constants, while
the z_k are the variables of the system (2).

For our example (see above) system (2) has the following form:

$$
\begin{aligned}
g_1(z_1) \quad &= z_1 - (x_1 + x_2) = 0 \\
g_2(z_1, z_2) \quad &= z_2 - z_1^3 \quad &= 0 \\
g_3(z_1, z_2, z_3) \quad &= z_3 - (x_3 - x_4) = 0 \\
g_4(z_1, z_2, z_3, z_4) \quad &= z_4 - z_2 * z_3 \quad &= 0 \\
g_5(z_1, z_2, z_3, z_4, z_5) &= z_5 * x_5 - z_4 \quad &= 0.
\end{aligned}
$$

System (2) will be solved by successive forward substitution which is equivalent
to the "normal" evaluation of a formula. Furthermore in our context, the partial
derivatives with respect to z_i have to exist for all g_k.

Let $\tilde{z}_1, \ldots, \tilde{z}_n$ be approximations for the actual solutions $\hat{z}_1, \ldots, \hat{z}_n$ of (2). In
the following we describe a possibility how to get inclusions for the errors

$$\Delta z_k := \hat{z}_k - \tilde{z}_k, \quad k = 1, \ldots, n. \tag{3}$$

We use the following abbreviations:

For $x, y \in D$, g_k from (2), $i \in \{1, \ldots, k\}$ and $\zeta \in x_i \cup y_i$ ($\cup$ denotes the convex hull
of the union) we define

$$
\begin{aligned}
r_i(g_k, x, y) &:= g_k(x_1, \ldots, x_{i-1}, y_i, \ldots, y_k) \\
s_i(g_k, x, y, \zeta) &:= \frac{\partial}{\partial z_i} g_k(x_1, \ldots, x_{i-1}, \zeta, y_{i+1}, \ldots, y_k).
\end{aligned}
\tag{4}
$$

Instead of $r_1(g_k, x, y)$ we write briefly $r(g_k, y)$. If $\dfrac{\partial}{\partial z_k} g_k$ is independent of z_k,
we may write $s_k(g_k, x, y, \zeta)$ as $s_k(g_k, x)$.

We apply the mean value theorem to the k-th equation of (2) with respect
to the k-th variable z_k; with (3) and (4) we get

$$g_k(\hat{z}_1, \ldots, \hat{z}_k) = r_k(g_k, \hat{z}, \tilde{z}) + s_k(g_k, \hat{z}, \tilde{z}, \zeta_k) \cdot \Delta z_k \quad (\zeta_k \in \hat{z}_k \cup \tilde{z}_k).$$

Repeating this step with respect to $z_j, j=k-1, k-2,\ldots$ yields

$$g_k(\hat{z}_1, \ldots, \hat{z}_k) = r(g_k, \tilde{z}) + \sum_{j=1}^{k} s_j(g_k, \hat{z}, \tilde{z}, \zeta_j) \cdot \varDelta z_j \qquad (\zeta_j \in \hat{z}_j \, \mathbf{U} \, \tilde{z}_j).$$

Since $\hat{z}_1, \ldots, \hat{z}_k$ is the exact solution of (2), the left-hand side of this expression is zero. For $\varDelta z_k$ we get the formula

$$\varDelta z_k = \left(-r(g_k, \tilde{z}) - \sum_{j=1}^{k-1} s_j(g_k, \hat{z}, \tilde{z}, \zeta_j) \cdot \varDelta z_j \right) \Big/ s_k(g_k, \hat{z}, \tilde{z}, \zeta_k) \tag{5}$$

provided that $s_k(g_k, \hat{z}, \tilde{z}, \zeta_k) \neq 0$.

For $k=1$ (5) reads as

$$\varDelta z_1 = -r(g_1, \tilde{z})/s_1(g_1, \hat{z}, \tilde{z}, \zeta_1). \tag{6}$$

If $s_1(g_1, \hat{z}, \tilde{z}, \zeta_1) = \dfrac{\partial}{\partial z_1} g_1(\zeta_1)$ is independent of ζ_1, then (6) is an appropriate formula for the computation of an inclusion $[\varDelta z_1]$ of $\varDelta z_1$. By means of induction it is easy to show that with inclusions $[\varDelta z_1], \ldots, [\varDelta z_{k-1}]$ for $\varDelta z_1, \ldots, \varDelta z_{k-1}$ the inclusion $[\varDelta z_k]$ of $\varDelta z_k$ can be computed with formula (5) if $s_k(g_k, \hat{z}, \tilde{z}, \zeta_k)$ is independent of z_k (i.e. g_k is at most linear in z_k). The ζ_j have to be substituted by $\hat{z}_j \, \mathbf{U} \, \tilde{z}_j$ and we take advantage of the already existing information

$$\hat{z}_j \in \tilde{z}_j + [\varDelta z_j], \quad j=1, \ldots, k-1.$$

We thus obtain the inclusion formula

$$\varDelta z_k \in [\varDelta z_k]$$
$$:= \left[-r(g_k, \tilde{z}) - \sum_{j=1}^{k-1} s_j(g_k, \tilde{z}+[\varDelta z], \tilde{z}, \tilde{z}_j \, \mathbf{U} \, (\tilde{z}_j+[\varDelta z_j])) \cdot [\varDelta z_j] \right] \Big/$$
$$\cdot s_k(g_k, \tilde{z}+[\varDelta z]), \tag{7}$$

where the right-hand side has to be computed using interval arithmetic.

The quality of an inclusion computed according to formula (7) highly depends on the accurate calculation of the residue term $r(g_k, \tilde{z})$. Because of transformation (2) the necessary accuracy can be easily achieved with the help of an exact scalar product (as to be seen below).

Moreover, the assumption that $\dfrac{\partial}{\partial z_k} g_k$ does not depend on z_k is always satisfied for that kind of equations g_k which arise from formula evaluation. This is shown in the next chapter.

2. Special Inclusion Formulae

Dealing with arithmetic expressions (including standard functions) only a few number of different types of equations g_k may occur. The different types and their corresponding inclusion formulae (7) are:

a) Addition

$$g_k(z_1, \ldots, z_k) = z_k - (z_i + z_j), \ i < j < k$$
$$[\Delta z_k] = -\tilde{z}_k + \tilde{z}_i + \tilde{z}_j + [\Delta z_i] + [\Delta z_j].$$

b) Subtraction

$$g_k(z_1, \ldots, z_k) = z_k - (z_i - z_j), \ i < j < k$$
$$[\Delta z_k] = -\tilde{z}_k + \tilde{z}_i - \tilde{z}_j + [\Delta z_i] - [\Delta z_j].$$

c) Multiplication

$$g_k(z_1, \ldots, z_k) = z_k - z_i \cdot z_j, \ i < j < k$$
$$[\Delta z_k] = -\tilde{z}_k + \tilde{z}_i \cdot \tilde{z}_j + \tilde{z}_j \cdot [\Delta z_i] + (\tilde{z}_i + [\Delta z_i]) \cdot [\Delta z_j].$$

d) Division

$$g_k(z_1, \ldots, z_k) = z_k \cdot z_j - z_i, \ i < j < k$$
$$[\Delta z_k] = (-\tilde{z}_k \cdot \tilde{z}_j + \tilde{z}_i - \tilde{z}_k \cdot [\Delta z_j] + [\Delta z_i])/(\tilde{z}_j + [\Delta z_j]).$$

e) Exponentiation (exponent is a positive integer)

$$g_k(z_1, \ldots, z_k) = z_k - z_i^n, \ i < k, \ n > 0$$
$$[\Delta z_k] = -\tilde{z}_k + \tilde{z}_i^n + n \cdot (\tilde{z}_i + ([\Delta z_i] \mathbf{U} \, 0))^{n-1} \cdot [\Delta z_i].$$

f) Exponentiation (exponent is a negative integer)

$$g_k(z_1, \ldots, z_k) = z_k \cdot z_i^{-n} - 1, \ i < k, \ n < 0$$
$$[\Delta z_k] = (-\tilde{z}_k \cdot \tilde{z}_i^{-n} + 1 + n \cdot \tilde{z}_k \cdot (\tilde{z}_i + ([\Delta z_i] \mathbf{U} \, 0))^{-n-1} \cdot [\Delta z_i])/(\tilde{z}_i + [\Delta z_i])^{-n}.$$

g) Standard functions

Let sf be a standard function (e.g. sin, cos etc.) and sf' its derivative:

$$g_k(z_1, \ldots, z_k) = z_k - sf(z_i), \ i < k$$
$$[\Delta z_k] = -\tilde{z}_k + sf(\tilde{z}_i) + sf'(\tilde{z}_i + ([\Delta z_i] \mathbf{U} \, 0)) \cdot [\Delta z_i].$$

In this case, the computation of the residue term $-\tilde{z}_k + sf(\tilde{z}_i)$ needs standard function procedures with dynamic accuracy and interval standard functions for the inclusion of sf'. Both are provided by [4], [7].

Formulae a)–d) can also be derived by simple algebraic transformations (with respect to the rules of interval analysis). All formulae will simplify once more if one (or both) of the operands is no intermediate result.

3. An Algorithm for the Evaluation of an Arithmetic Expression

As indicated in chapter 2, the idea of the computation of bounds for the value of a formula is essentially based on two steps:

(a) Computation of an approximation for all intermediate results $\tilde{z}_k$ ($k = 1, \ldots, n$);

(b) Computation of inclusions $[\Delta z_k]$ for the errors Δz_k ($k = 1, \ldots, n$).

If the equality of the result $\tilde{z}_n + [\varDelta z_n]$ is not sufficient, the midpoint $m([\varDelta z_k])$ of $[\varDelta z_k]$ $(k = 1, \ldots, n)$ will be used to improve the approximation $\tilde{z}_k$ to

$$\tilde{z}_k + m([\varDelta z_k]). \tag{8}$$

Then step (b) is started again and so on.

The improvement of the approximation, however, is not done by adding $\tilde{z}_k$ and $m([\varDelta z_k])$ explicitly, but by storing $\tilde{z}_k$ and $m([\varDelta z_k])$ separately in a correction vector.

After the first step we set: $z_k^{(0)} := \tilde{z}_k$, $z_k^{(1)} := m([\varDelta z_k])$ and $[\varDelta z_k^{(1)}] := [\varDelta z_k]$. In the r-th correction step the approximation $\tilde{z}_k$ looks like

$$\tilde{z}_k = \sum_{s=0}^{r} z_k^{(s)}.$$

Of course, this has to be taken into account in all formulae of chapter 2. E.g. in the r-th step instead of formula a) the following has to be computed:

$$[z_k^{(r)}] := \Diamond \left[-\sum_{s=0}^{r-1} z_k^{(s)} + \sum_{s=0}^{r-1} z_i^{(s)} + \sum_{s=0}^{r-1} z_j^{(s)} \right] \Leftrightarrow [\varDelta z_i^{(r)}] \Leftrightarrow [\varDelta z_j^{(r)}].$$

By $\Diamond r$ $(r \in \mathbb{R})$ we denote an including interval with computer representable bounds that contains r, by $\Leftrightarrow$ computer interval addition in the sense of [8]. Obviously, the residue term may be interpreted as a scalar product.

If the residue term of the $(r-1)$-th step, i.e. $R_k^{r-1} := -\sum_{s=0}^{r-2} z_k^{(s)} + \sum_{s=0}^{r-2} z_i^{(s)} + \sum_{s=0}^{r-2} z_j^{(s)}$

in our example, has been stored, then R_k^r can be computed by

$$R_k^r = R_k^{r-1} - z_k^{(r-1)} + z_i^{(r-1)} + z_j^{(r-1)},$$

thus saving computation time by using some extra storage. The cases b)–d) can be treated analogously.

For interval parameters a slightly different approach is necessary. The interested reader is refered to an article in [6]. There, the treatment of interval input by a subdivision method [9] is described in some detail.

The whole algorithm is displayed as Algorithm 1. We add the following remarks: The first termination criterion applies to the case where the desired accuracy has been achieved. With the assumption that $\delta = B^{-t}$, where B is the radix and t the mantissa length of the machine numbers, a result is expected whose upper and lower bounds differ only in the last digit [3].

The second criterion is used for small-tolerances-afflicted input data. It may occur that the diameters of these intervals prevent a further improvement of the result.

Algorithm 1

1. Compute approximations $z_k^{(0)}$ of all intermediate results $(k = 1, \ldots, n)$;

2. {Iteration}
   $r := 0$;
   repeat
   $r := r + 1$;
   if $r > 1$ **then** $z_k^{(r-1)} := m([z_k^{(r-1)}])\ (k = 1, \ldots, n)$;
   Compute inclusions $[z_k^{(r)}]\ (k = 1, \ldots, n)$ according to the formula from chapter 2;

   $$[y^{(r)}] := \Diamond\left(\sum_{j=0}^{r-1} z_n^{(j)}\right) \Diamond [z_n^{(r)}];\ \{\text{formula value}\}$$

   until $(d([z_n^{(r)}]) \le \delta |[y^{(r)}]|)$
   $\{d(A)$ denotes the diameter of interval $A\}$
   or $([y^{(r)}] = [y^{(r-1)}])$
   or (underflow **and** $r \ge 10$);

At last, the third criterion is necessary if during computation an underflow occurs. This may indicate that an intermediate result needs more digits than those given by the actual floating-point numbers. If this happens, the algorithm does not terminate until 10 corrections have been tried. It turns out that 10 is a useful value for a domain of exponents reaching from -99 to $+99$.

With a view to briefing, we will not go into further details such as overflow handling or avoiding divisions by zero. We only mention here that these exceptions are treated by a recursive call of the formula evaluation procedure for critical subterms.

Within the framework of the ESPRIT project DIAMOND [5] Algorithm 1 was implemented in the PASCAL extension PASCAL-SC [2], [12], [13].

4. Extension for Matrix Expressions and Complex Expressions

The strategy of chapter 2 can be extended in a natural way to the evaluation of matrix- and complex expressions.

For simple matrix expressions (consisting of $+$, $-$, $*$, $/$) the error formulae can be derived analogously in a pure algebraic way. Of course, the non-commutativity of matrix multiplication has to be taken into account.

For matrix exponentiation the defect formulae 2e) and f) are replaced by

e') $\quad [\Delta Z_k] = -\tilde{Z}_k + \tilde{Z}_i^m + \sum\limits_{j=0}^{m-1} \tilde{Z}_i^j * [\Delta Z_i] * (\tilde{Z}_i + [\Delta Z_i])^{m-1-j}.$

(Since $(\tilde{Z}_i + \Delta Z_i)^m = \tilde{Z}_i^m + \sum\limits_{j=0}^{m-1} \tilde{Z}_i^j * \Delta Z_i * (\tilde{Z}_i + \Delta Z_i)^{m-1-j}$, what can be proved by induction.)

f') $\quad [\Delta Z_k] = \left(1 - \tilde{Z}_k * \tilde{Z}_i^m - \sum\limits_{j=0}^{m-1} \tilde{Z}_k * \tilde{Z}_i^j [\Delta Z_i](\tilde{Z}_i + [\Delta Z_i])^{m-1-j}\right) / (\tilde{Z}_i + [\Delta Z_i])^m.$

For complex expressions the defect formulae 2a)—g) are valid literally. For their deduction the mean value theorem applies only componentwise for the real and imaginary part. However, the following theorem is valid [3]:

Let $Z \subset \mathbb{C}$, $\tilde{z} \in \mathbb{C}$, $G \supset Z \cup \tilde{z}$ a domain, and
$f: G \rightarrow \mathbb{C}$ holomorphic. Then
$f(Z) \subset f(\tilde{z}) + [f'(Z \cup \tilde{z}) \cdot (Z - \tilde{z})].$

The residue terms can still be computed by scalar products. For the implementation of accurate complex interval standard functions see [4], [7].

5. Example

The following expression [10] demonstrates the fact that even large mainframes may not only compute wrong results but also suggest an accuracy which does not exist:

$$333.75 \cdot b^6 + a^2 \cdot (11 \cdot a^2 \cdot b^2 - b^6 - 121 \cdot b^4 - 2) + 5.5 \cdot b^8 + a/(2 \cdot b).$$

Exponentiation was done by repeated multiplication.

On an IBM 4361 using a simple FORTRAN program with $a = 77617.0$ and $b = 33096.0$ the following results (rounded to 7 decimal digits) were computed:

real	(6 hex digits):	1.172604
double precision	(14 hex digits):	1.172604
extended precision	(28 hex digits):	1.172604.

Thus, even though the results suggest a fairly harmless problem, the correct result lies in the interval $[-0.8273961, -0.8273960]$, which was computed with an implementation of Algorithm 1 in PASCAL-SC.

The corresponding inclusions $[y^1]$, $[y^2]$ and $[y^3]$ of the final intermediate result in Algorithm 1 demonstrate the ill condition of the problem (naturally this can be shown with any interval arithmetic) and the typical convergence of the method:

$[y^1] = [-1.1 \text{E} 12, 4.0 \text{E} 12]$
$[y^2] = [-2.0, 0.0]$
$[y^3] = [-0.827\,396\,059\,946\,9, -0.827\,396\,059\,946\,7].$

Acknowledgement

The authors are grateful to A. Neumaier for formula 4e'). The formula we used before gave slightly weaker inclusions and had to be proved by a generalized mean value theorem.

References

[1] Alefeld, G., Herzberger, J.: Introduction to Interval Computations. Academic Press, New York, 1983.

[2] Bohlender, G., Rall, L. B., Ullrich, Ch., Wolff von Gudenberg, J.: PASCAL-SC: A Computer Language for Scientific Computation. Academic Press (Perspectives in Computing, vol. 17), Orlando, 1987.

[3] Böhm, H.: Berechnung von Polynomnullstellen und Auswertung arithmetischer Ausdrücke mit garantierter maximaler Genauigkeit, Dissertation, Universität Karlsruhe, 1983.

[4] Braune, K.: Standardfunktionen für reelle und komplexe Punkt- und Intervallargumente mit dynamischer Genauigkeit, Dissertation, Universität Karlsruhe, 1987 (see also this volume).

[5] Fischer, H. C., Haggenmüller, R., Schumacher, G.: Evaluation of Arithmetic Expressions, DIA-MOND, Deliverable D 2a – 1, Doc. No.: 03/2a – 1/1K02.f.

[6] Fischer, H. C., Haggenmüller, R., Schumacher, G.: Evaluation of Arithmetic Expressions with Guaranteed High Accuracy, Siemens Forschungs- und Entwicklungsberichte, No. 5, 1987, pp. 171 – 177.

[7] Krämer, W.: Inverse Standardfunktionen für reelle und komplexe Punkt- und Intervallargumente mit dynamischer Genauigkeit, Dissertation, Universität Karlsruhe 1987 (see also this volume).

[8] Kulisch, U., Miranker, W. L.: Computer Arithmetic in Theory and Practice. Academic Press, New York, 1981.

[9] Ratschek, H., Rokne, J.: Computer Methods for the Range of Functions. Ellis Horwood, Chichester, 1984.

[10] IBM High-Accuracy Arithmetic Subroutine Library (ACRITH). Program Description and User's Guide. SC 33-6164-02, 3 rd Edition, April 1986.

[11] ARITHMOS Benutzerhandbuch. SIEMENS AG, Bestellnummer U 2900-J-Z 87-1, Sept. 1986.

[12] Kulisch, U. (ed.): PASCAL-SC: A PASCAL extension for scientific computation; information manual and floppy disks; version IBM PC/AT; operating system DOS. B. G. Teubner Verlag (Wiley-Teubner series in computer science), Stuttgart, 1987.

[13] Kulisch, U. (ed.): PASCAL-SC: A PASCAL extension for scientific computation; information manual and floppy disks; version ATARI ST. B. G. Teubner Verlag, Stuttgart, 1987.

Dipl. Math. Hans Christoph Fischer
Institut für Angewandte Mathematik
Universität Karlsruhe
Kaiserstrasse 12
D-7500 Karlsruhe
Federal Republic of Germany

Priv.-Doz. Dr. rer. nat. habil. Rudolf Haggenmüller
Siemens AG München
Geschäftsbereich Datentechnik
Otto-Hahn-Ring 6
D-8000 München 83
Federal Republic of Germany

Dipl. Math. Günter Schumacher
Institut für Angewandte Mathematik
Universität Karlsruhe
Kaiserstrasse 12
D-7500 Karlsruhe
Federal Republic of Germany

Computing, Suppl. 6, 159–184 (1988)

Standard Functions for Real and Complex
Point and Interval Arguments with Dynamic Accuracy[1]

Klaus Braune, Karlsruhe

Abstract – Zusammenfassung

Standard Functions for Real and Complex Point and Interval Arguments with Dynamic Accuracy.
Algorithms for the high-accuracy evaluation of the real and complex standard functions e^x, $\sin x$,
$\cos x$, $\tan x$, $\cot x$, $\sinh x$, $\cosh x$, $\tanh x$, $\coth x$, $\sqrt{x}$ and x^2 for point and interval arguments in general
floating-point screens are treated. In case of interval functions an inclusion of the exact result is
computed. The number of guard digits necessary to obtain this result in general floating-point screens
is specified for each function. The error bounds stated in this paper are proved in [3]. The missing
inverse standard functions can be found in short in [6] or more detailed in [7].

Standardfunktionen für reelle und komplexe Punkt- und Intervallargumente mit dynamischer Genauigkeit. Es werden Algorithmen zur hochgenauen Berechnung der reellen und komplexen Standardfunktionen e^x, $\sin x$, $\cos x$, $\tan x$, $\cot x$, $\sinh x$, $\cosh x$, $\tanh x$, $\coth x$, $\sqrt{x}$ und x^2 für Punkt- und
Intervallargumente in beliebigen Gleitpunktrastern mit Fehlerabschätzungen angegeben. Im Fall
der Intervall-Funktionen wird das exakte Intervall eingeschlossen. Die Anzahl der zur Berechnung
benötigten Schutzziffern für die einzelnen Funktionen in beliebigen Gleitpunktrastern wird in einer
Tabelle angegeben. Die in diesem Artikel angegebenen Fehlerschranken werden in [3] hergeleitet.
Die fehlenden inversen Standardfunktionen werden in [6] kurz diskutiert oder ausführlicher in [7].

1 Introduction

The standard functions discussed in this paper and in Krämer [6] have been
implemented for 2 specific floating-point data formats as part of the IBM
program product ACRITH [1]. The mathematical problems occurring in this
implementation have been presented in [4]. The definitions of the terms used in
this paper like real and complex *interval functions*, *branch cut* and *many-valued
functions* can be found in [4] together with illustrating figures.

These methods in [3] and [7] are extended to arbitrary floating-point data formats
including rigorous error bounds.

In this paper the *implementation* of the standard functions for arbitrary data
formats is treated. Bounds for the error when computing the function value using
the specified algorithm are given. All given formulas are proved in [3] and [7].

---

[1] Diese Arbeit wurde mit Mitteln des Ministeriums für Wirtschaft, Mittelstand und Technologie
Baden-Württemberg gefördert.

1.1 Motivation

Standard functions are an integral part of modern computer arithmetic as well as of all computer languages designed for numerical computation. Results of numerical computations are useful only if their accuracy is (at least approximately) known. Hence, it is not only neccessary to take into account the error due to the algorithm (which is usually known), but also to keep track of *all* rounding errors occurring in every operation due to the fact that floating-point numbers are used, not real or complex numbers. Obviously, the error of each floating-point operation — and of the standard functions — must be known. However, standard functions are commonly supplied without error bounds. Thus, algorithms for the computation of real and complex standard functions *with rigorous error bounds* are needed.

Since the rounding errors depend on the computer the standard functions are designed for, each implementation requires different error estimates. On the other hand, rigorous error bounds can be proved nearly independently of the target machine — with minor restrictions. This allows an implementation of the standard functions with rigorous error bounds for different machines and floating-point data formats without repeating the error estimates. This holds for *real* and *complex* functions.

As long as mechanical tools were used, the rounding errors could be controlled manually. Since the development of modern computers, algorithms with millions of arithmetic operations are often used. It is impossible to manually keep track of the rounding errors occurring in each operation performed. However, interval arithmetic allows to keep track of the rounding errors *automatically* as well as handling problems with disturbed data or interval problems. Obviously, arithmetical operations for intervals and interval standard functions are necessary. However, application of naive interval methods may result in a great overestimation of rounding errors which may even cause the algorithm to fail. Self-validating methods as discussed in this volume can often be used to reduce these errors and to compute an inclusion of the true result. Among other things, these methods require interval standard functions.

Algorithms for highly accurate interval standard functions must return a result containing *all* exact results for each point in the argument interval. Monotonic functions merely require evaluation at the bounds and then decreasing the lower bound and increasing the upper bound of the result by a rigorous error bound. The trigonometric real functions and most of the complex functions require additional investigation. Some of the problems occurring when high-accuracy interval standard functions are implemented are shown in [4].

1.2 Notations

The functions under consideration are designed for a screen

$$\mathscr{S}^0 = \mathscr{S}(B, \ell^0, em^0, eM^0).$$

Computation of high-accuracy results requires some guard digits. All arithmetic operations are executed with these guard digits, i.e. in a *finer* screen

$$\mathscr{S} = \mathscr{S}(B, \ell, em, eM)$$

with base B, minimal exponent em, maximal exponent eM and ℓ mantissa digits. These quantities are related to the corresponding quantities of the screen $\mathscr{S}^0$ according to

$$\ell \geq \ell^0 + k, \quad em \leq em^0 \text{ and } eM \geq eM^0,$$

where $k > 0$ is chosen according to tables 11 and 12. The arithmetic of the screen $\mathscr{S}^0$ is not used. Hence, only the screen $\mathscr{S}$ is concerned with arithmetical requirements. The screen operations are marked by a circle $\bigcirc$ in contrast to *exact* operations.

Intervals are specified as $[\underline{x}, \overline{x}]$ with $\underline{x}$ being the lower bound and $\overline{x}$ being the upper bound.

Complex numbers are represented in the form $z = x + iy$ using the real part $\operatorname{Re} z = x$ and the imaginary part $\operatorname{Im} z = y$. A *complex interval* $[\underline{z}, \overline{z}]$ is a rectangle in the complex plane with $\underline{z}$ being the lower left corner and $\overline{z}$ being the upper right corner.

Rounding errors are denoted by ε. A subscript marks a special error term, the error bound for the function f is specified as $\varepsilon(f)$. If neither argument nor subscript are present, ε means an error of 1 ulp for the screen $\mathscr{S}$. The term $\varepsilon(\ell - m)$ represents the corresponding error for a screen with $\ell - m$ digits, i.e. a loss of m digits.

Directed roundings are marked by $\triangle$ (upwards) and $\triangledown$ (downwards).

1.3   Arithmetical Requirements

On $\mathscr{S}$ the operations $\{+, -, \cdot, /\}$ are required. The relative error of the screen operations is assumed to be bounded according to

$$\left| \frac{x \otimes y - x \times y}{x \times y} \right| < \varepsilon := B^{1-\ell}.$$

ℓ may be less than the number of mantissa digits actually present.

For some special purposes an *exact product* of screen numbers is necessary. It may be computed by splitting the mantissa into parts, multiplying the different parts and accumulating the products. For all cases appearing in this paper the accumulation can be performed using operations of the screen $\mathscr{S}$.

Multiplication by powers of the base and addition of 0 are assumed to be *exact*.

In case of interval functions, directed roundings from the screen $\mathscr{S} \to \mathscr{S}^0$ are necessary. Since $\mathscr{S}$ is finer than the screen $\mathscr{S}^0$, directed roundings are assumed to yield the nearest number of the screen $\mathscr{S}^0$ in the specified direction. In case of pointwise functions, the rounding is assumed to yield one of the neighbouring points of the screen $\mathscr{S}^0$.

All constants are assumed to be available as numbers of the screen $\mathscr{S}$ with relative error less than ε, if not otherwise specified.

1.4 Approximations

Computation of function values of almost all standard functions requires an approximation, since all representations need an infinite number of arithmetical operations. When computing standard functions with dynamic accuracy, using the leading terms of their Taylor series seems to be a good choice. Alternatively, continued fractions could be used, but in many cases division is a relatively slow operation. Hence, Taylor polynomials are preferred. For all treated functions the approximation error $\varepsilon(app)$ when using up to 20 leading terms of the Taylor series is tabulated in [3]. The specified total error bounds are based on the evaluation of the approximation polynomial according to Horner's scheme. The bounds are valid, if the degree N of the Taylor polynomial satisfies

$$N \le \frac{1}{2} \cdot \left(\sqrt{\frac{2}{\varepsilon}} - 1 \right).$$

1.5 Interval Evaluation

Real intervals are specified by giving their lower and upper bounds. The complex interval standard functions are designed for complex intervals being rectangles in the complex plane. Thus, real and imaginary parts of complex intervals are real intervals.

The evaluation of a function with an interval argument is based on the evaluation of this function at appropriate points. In case of monotonic functions $f(x)$, these points are the lower and upper bound of the argument:

$$f([\underline{x}, \overline{x}]) = \begin{cases} [f(\underline{x}), f(\overline{x})], & f \text{ increasing}, \\ [f(\overline{x}), f(\underline{x})], & f \text{ decreasing}. \end{cases} \tag{1}$$

Otherwise, special considerations are necessary to find the points for evaluation. If a function is not monotonic, these points are specified with the function.

In case of complex interval standard functions both the real and the imaginary part are handled as real interval functions depending on 2 real interval arguments. The points where the real and the imaginary part attain its lower and upper bounds vary for different arguments. Hence, these points are specified using tables.

In addition, the errors of the computed pointwise function values have to be taken into account by multiplying the computed value by an error term $1 \pm 1.1 \cdot (\varepsilon(f) + \varepsilon)$ compensating the rounding error of the multiplication, too. In case of a negative lower bound or a positive upper bound the factor $1 + 1.1 \cdot (|\varepsilon(f)| + |\varepsilon|) > 1$ is to be used, in case of a positive lower bound or a negative upper bound the factor $1 - 1.1 \cdot (|\varepsilon(f)| + |\varepsilon|) < 1$ is to be used.

Finally, the computed lower bound and upper bound have to be rounded downwards and upwards into the screen $\mathscr{S}^0$, respectively.

2   Real Functions

2.1   Exponential Function

Range. — The function value of the exponential function is representable in the screen $\mathscr{S}$ when

$$0.5 \cdot B^{em} \le e^x < B^{eM} \iff em \cdot \ln B - \ln 2 \le x < eM \cdot \ln B.$$

Otherwise overflow or underflow occurs.

Approximation. — The series expansion of the exponential function is

$$e^x = \sum_{n=0}^{\infty} \frac{x^n}{n!},$$

which converges for all real x.

The error when approximating the function value using the leading $N + 1$ terms of the Taylor series for $x \ge 0$ is bounded by

$$e^{-x} \cdot \left| e^x - \sum_{n=0}^{N} \frac{x^n}{n!} \right| \le \frac{1}{(N+1)!} \cdot x^{N+1}. \tag{2}$$

(2) is used to determine an appropriate value for N. To avoid the evaluation of polynomials of high degree, the series is assumed to be used exclusively within the interval $[0, 0.1]$.

Argument Reduction. — An argument reduction is necessary to match the condition above. The formula used is

$$e^x = B^{B^{-n} \cdot [B^n \cdot x / \ln B]} \cdot e^{x - B^{-n} \cdot \ln B \cdot [B^n \cdot x / \ln B]}.$$

The exponent of e obviously lies within $[0, B^{-n} \cdot \ln B)$, and the exponent of B is an integer multiple of B^{-n} and hence representable using the base B with at most n fractional digits:

$$B^{-n} \cdot [B^n \cdot x / \ln B] = [B^{-n} \cdot [B^n \cdot x / \ln B]] + \sum_{m=1}^{n} \beta_m \cdot B^{-m}.$$

The integer part requires no further consideration. The fractional part is taken into account by the B^n constants

$$c(\beta_1, \beta_2, \ldots, \beta_n) = \exp\left(\ln B \cdot \sum_{m=1}^{n} \beta_m B^{-m} \right), \quad \beta_m \in \mathbb{N}, \; 0 \le \beta_m \le B - 1.$$

The number of required constants may be reduced by replacing k subsequent constants by a single constant as long as the interval $[0, k \cdot B^{-n} \cdot \ln B)$ is part of the interval $[0, 0.1)$. In this case the index I of the constant c and the reduced argument t are computed according to

$$j = B^n \cdot \sum_{m=1}^{n} \beta_m B^{-m},$$

$$t = x - B^{-n} \ln B \cdot \{[B^n x / \ln B] + (j \bmod k)\},$$
$$I = j//k \qquad (j \text{ div } k).$$

The needed integer part is the product of the absolute value of the argument x and the constant

$$\widetilde{F} = \triangle_\ell \{B^n \oslash \ln B \odot (1 + 1.1\varepsilon)\},$$

yielding the approximation

$$j^0 = [|x| \odot \widetilde{F}],$$

which is either j or $j + 1$ according to [3].

Error Bound. — The total relative error of the computed function value is bounded by

$$\varepsilon(\exp) = \; 2.01 \ln B \cdot \varepsilon(\ell + n - 1) + 1.01 \cdot (|x|/ \ln B + B^{-n}) \cdot \varepsilon(k - n - 1) \qquad (3)$$
$$+ \varepsilon(app) + 4.31\varepsilon + 50 \max {}^2\{\varepsilon, \varepsilon(app), \varepsilon_t\}$$

where $k \gg n$ is the number of mantissa digits used of F^{-1} and the approximation error $\varepsilon(app)$ is given by (2). ε_t is an abbreviation of the first line of the right-hand side in (3).

Algorithm. — The value $\widetilde{\exp}(x)$ is computed according to the following steps:

1. Compute the integer part $j = [|x| \odot \widetilde{F}]$.

2. Compute the reduced argument by splitting F^{-1} into parts $ip_i, i = 1 \ldots K$, which can be multiplied exactly by j :

   $y := |x|; \quad i := 1;$
   Repeat $y := y - j \cdot ip_i; \quad i := i + 1;$
       Until $(i > K \; \underline{Or} \; "No \; leading \; digit \; lost");$
   $t := 0;$
   For $k := K$ *Downto* i *Do* $t := t \oplus j \cdot ip_k;$
   $y := y \ominus t;$

   If $y < 0$ diminish j by 1 and repeat this step.

3. Compute an approximation $\widetilde{\exp}(t)$ using Horner's scheme and the function value

$$\widetilde{\exp}(x) := \begin{cases} B^{j//B^n} \cdot \widetilde{c}_{j \bmod B^n} \odot \widetilde{\exp}(t) & x \geq 0, \\ B^{-j//B^n} \oslash (\widetilde{c}_{j \bmod B^n} \odot \widetilde{\exp}(t)) & \text{otherwise.} \end{cases}$$

The function value of the exponential function for a real argument and result is found by rounding the computed function value into the screen $\mathscr{S}^0$. Since e^x is strictly increasing, the interval evaluation is obvious.

2.2   Sine and Cosine

Range. —   The function value of sine and cosine is representable for all arguments. However, to perform a correct argument reduction, the range should be restricted.

Approximation. —   The series expansion of the sine function is

$$\sin x = \sum_{n=0}^{\infty} (-1)^n \cdot \frac{x^{2n+1}}{(2n+1)!},$$

which converges for all real x. However, not the function $\sin x$ is approximated, but $\sin \frac{\pi}{2} t$.

The cosine is computed according to

$$\cos x = \sin \left(x + \frac{\pi}{2} \right).$$

The error when approximating the function value using the leading $N + 1$ terms of the Taylor series for $t \in [-1, 1]$ is bounded by

$$\frac{1}{|\sin \frac{\pi t}{2}|} \cdot \left| \sin \left(\frac{\pi t}{2} \right) - \sum_{n=0}^{N} (-1)^n \cdot \frac{\left(\frac{\pi}{2} \cdot t \right)^{2n+1}}{(2n+1)!} \right| \leq \left(\frac{\pi}{2} \right)^{2N+3} \cdot \frac{t^{2n+2}}{(N+1)!}. \tag{4}$$

Using (4), an appropriate N can be determined.

Argument Reduction. —   An argument reduction is necessary to prevent the evaluation of polynomials of a high degree and cancellation of leading digits. The argument is reduced by splitting off integer multiples of $\pi/2$. First, the argument is multiplied by $2/\pi$. Usually, relatively few digits of $2/\pi$ are adequate. However, if the reduced argument is close to 0, additional digits have to be used. This fact can be checked *after* performing the reduction with few digits. If necessary, the accuracy can be improved dynamically.

The algorithm specified below works if the following assumptions hold:

1. The value $2/\pi$ is available with a sufficient number of digits as an array of screen numbers, which can be multiplied *exactly* with the specified argument. It may be necessary to split the argument into 2 parts to satisfy this requirement.

2. The products are stored as an array of screen numbers with exponents differing exactly by ℓ.

3. The result of an addition or subtraction must be *exact* if leading digits are cancelled.

The reduced argument is computed according to

$$j := \left[x \cdot \frac{2}{\pi} \right] \tag{5}$$

$$t := \begin{cases} \frac{2}{\pi} \cdot x - j, & j \bmod 4 = 0 \\ (j+1) - \frac{2}{\pi} \cdot x, & j \bmod 4 = 1 \\ j - \frac{2}{\pi} \cdot x, & j \bmod 4 = 2 \\ \frac{2}{\pi} \cdot x - (j+1), & j \bmod 4 = 3 \end{cases} \tag{6}$$

The integer part j of $x \cdot 2/\pi$ is used to select the appropriate formula in (6). In case of an interval argument, the bounds of the result depend on j as well.

Error Bound. — The error bound below is specified assuming that for the computed reduced argument $\tilde{t} = \widetilde{mt} \cdot B^{\widetilde{et}}$ the inequality

$$\widetilde{et} \geq 1 + \ell + ex - k \tag{7}$$

holds, where $x = mx \cdot B^{ex}$. Replacing $\widetilde{et}$ by the exponent em of the smallest number treated as being different from 0 shows that in no case more than

$$k \geq 1 + \ell + ex - em \tag{8}$$

digits of $2/\pi$ are required.

The total error of the computed value $\sin x$ or $\cos x$ is bounded by

$$\varepsilon(\sin) = \varepsilon(\ell - 1) \cdot \left(1.03 + \frac{18.7}{B} \right) + \varepsilon(app) + \varepsilon(\ell - 1) \cdot (11\varepsilon(app) + 41\varepsilon(\ell - 1)) \tag{9}$$

where the approximation error $\varepsilon(app)$ is given by (4).

Algorithm. — The computation of $\widetilde{\sin}(x)$ is based on the data

    Argument : $x \in \mathscr{S}$
                    Initialization of the integer part j (0 for sine, 1 for cosine)
                    Exponent U of the smallest result treated $\neq 0$
    Result   : $\widetilde{\sin}(x) \in \mathscr{S}$
                    Integer part $j \in \mathbb{N}$

The algorithm consists of the following steps:

1. Multiply *exactly* $|x|$ by the first unused part ip_v with m mantissa digits and increase the number k of digits used by m. Add the integer part of the product to j and store the fractional part as v_v.

Sum up the v_μ in reverse order (starting with v_0). If the sum exceeds 1, increase j by 1 and add the terms v_μ successively to -1 starting with v_0. As long as the sum is negative, set the corresponding v_μ to 0. If the sum becomes positive, replace the corresponding v_μ by the sum itself.

Compute the reduced argument t according to (6): If j is even, sum up the v_μ in reverse order, if j is odd, subtract the fractional part from 1 beginning with v_0.

The argument reduction is finished if $\widetilde{et} < U$ or if (7) holds. Otherwise repeat this step.

2. If $j \bmod 4 > 1$, replace t by $-t$.

3. In case $x < 0$ the integer part j has to be modified according to

$$j := 7 - (j - 2j_0),$$

where j_0 is the initial value of j.

4. Compute an approximation of $\widetilde{\sin \frac{\pi t}{2}}$ using Horner's scheme and return the result as well as $j \bmod 8$.

Finally, for a real argument and result, the computed function value of the sine (or cosine) function is rounded into the screen $\mathscr{S}^0$.

Real Interval Arguments. — In case $\overline{x} - \underline{x} \geq 2\pi$ the result interval obviously is $[-1, 1]$.

Otherwise, the bounds of the result depend on the integer parts $\underline{j}$ and $\overline{j}$ of the bounds $\underline{x}$ and $\overline{x}$ (modulo 8) of the argument interval.

In case of $\underline{j} = \overline{j}$, $\sin x$ is monotic within the argument interval. The result interval is

$$\sin([\underline{x}, \overline{x}]) = \begin{cases} [\sin \underline{x}, \sin \overline{x}], & \underline{j} \bmod 4 = 0, 3 \\ [\sin \overline{x}, \sin \underline{x}], & \underline{j} \bmod 4 = 1, 2 \end{cases} \qquad (10)$$

Otherwise, the bounds have to be chosen appropriately. With

$$j = \underline{j} \bmod 4, \qquad d = (\overline{j} - \underline{j}) \bmod 4,$$
$$\text{MIN} = \min\{\sin \underline{x}, \sin \overline{x}\}, \qquad \text{MAX} = \max\{\sin \underline{x}, \sin \overline{x}\},$$

the result interval is shown in table . Using this table, the implementation of an algorithm for the interval evaluation of $\sin x$ and $\cos x$ is obvious.

j \ d	1	2	3
0	[MIN, 1]	[MIN, 1]	[-1 , 1]
1	[MIN,MAX]	[-1 ,MAX]	[-1 ,MAX]
2	[-1 ,MAX]	[-1 ,MAX]	[-1 , 1]
3	[MIN,MAX]	[MIN, 1]	[MIN, 1]

Table 1: Result intervals for the interval sine function.

2.3 Tangent and Cotangent

Range. — The tangent and cotangent functions are defined for all *real* arguments differing from the poles $\pi/2 + k\pi$ in case of $\tan x$ and $k\pi$ in case of $\cot x$. However, the poles are not screen numbers except 0 for the cotangent. In general, not even overflow should occur for screen numbers, but this case cannot be excluded.

Point Arguments. — The tangent and cotangent functions are computed using the sine and cosine functions according to

$$\tan x = \frac{\sin x}{\cos x}, \qquad \cot x = \frac{\cos x}{\sin x}.$$

For both functions the relative rounding error is bounded by

$$\varepsilon(\tan) = 2\varepsilon(\sin) + \varepsilon + 4.4\varepsilon(\sin)^2. \tag{11}$$

The algorithm for computation is obvious. However, before dividing sine and cosine, the possibility of an overflow has to be checked.

Real Interval Arguments. — The tangent is strictly increasing and the cotangent strictly decreasing between the poles. The interval evaluation is evident if there is no pole within the argument interval.

A pole is located within the argument interval if the integer parts j and $\bar{j}$ are not equal and j is *even* for the tangent and *odd* for the cotangent, or if the bounds $\underline{x}$ and $\bar{x}$ differ by at least π.

2.4 Hyperbolic Sine

Range. — The hyperbolic sine function is defined for all real arguments. Because of $|\sinh x| \approx 0.5 \cdot e^x$, the restrictions for the exponential function with respect to a representable result hold for this function, too. For large $|x|$ the function value is computed by

$$\sinh x = \operatorname{sgn} x \cdot \left(e^{|x|} - e^{-|x|} \right) / 2. \tag{12}$$

This formula can be used as long as no leading digits are cancelled by the subtraction of the exponential terms. Otherwise, i.e. for x close to 0, an approximation has to be used.

Approximation. — Cancellation of leading digits of $e^{|x|}$ is impossible in case $|\sinh x| \geq 0.5$. Hence, the approximation can be restricted to the range $|x| < 0.4813$. The function value is approximated using the leading terms of the Taylor series

$$\sinh x = x \cdot \sum_{n=0}^{\infty} \frac{x^{2n}}{(2n+1)!}$$

which converges for all real x.

The relative approximation error when using the leading $N + 1$ terms is bounded by

$$\frac{1}{|\sinh x|} \cdot \left| \sinh x - x \cdot \sum_{n=0}^{N} \frac{x^{2n}}{(2n+1)!} \right| \leq \frac{x^{2N+2}}{(2N+3)!} \cdot \cosh x. \tag{13}$$

From (13) an appropriate value of N can be obtained.

Error Bound. — Combining the preceding error estimates with the rounding errors in the operations to be performed yields the bound

$$\varepsilon(\sinh) = \max \; \{ \varepsilon(\exp) + 2.59\varepsilon + 4.4\varepsilon(\exp)^2, \tag{14}$$
$$\varepsilon(app) + 2.17\varepsilon + 2.2\varepsilon(app) \cdot \varepsilon \}$$

for the relative error when computing the hyperbolic sine function, where the approximation error $\varepsilon(app)$ is given by (13) and $\varepsilon(\exp)$ is given by (3).

Algorithm. — The steps for the computation of $\widetilde{\sinh}(x)$ are

1. If $|x| \geq 0.4813$, compute $\sinh x$ according to (12).

2. If $|x| < 0.4813$, compute an approximation value by evaluating the Taylor polynomial of appropriate degree according to Horner's scheme.

Since $\sinh x$ is strictly increasing, the algorithm for interval evaluation is evident.

2.5 Hyperbolic Cosine

Range. — For the range of the hyperbolic cosine the same restriction holds as for the hyperbolic sine. However, since the hyperbolic cosine is defined by

$$\cosh x = \left(e^x + e^{-x} \right) / 2, \tag{15}$$

there is no danger of cancellation of leading digits. Thus, the given formula can be used for all arguments.

Error Bound. — The relative error in computing $\cosh x$ using the above formula is bounded by

$$\varepsilon(\cosh) = \varepsilon(\exp) + 2.51 \cdot \varepsilon + 4.4 \cdot \varepsilon(\exp)^2, \tag{16}$$

where $\varepsilon(\exp)$ is given by (3).

The computation of $\widetilde{\cosh}(x)$ is performed according to (15) using the exponential function.

Real Interval Arguments. — The hyperbolic cosine is an even function with minimum at $x = 0$, i.e. for negative arguments $\cosh x$ is monotonically decreasing and for positive arguments monotonically increasing. Thus, the interval evaluation yields the result

$$\cosh([\underline{x}, \overline{x}]) = \begin{cases} [\cosh \underline{x}, \cosh \overline{x}], & \underline{x} \geq 0, \\ [\cosh \overline{x}, \cosh \underline{x}], & \overline{x} \leq 0, \\ [1, \cosh(\max\{-\underline{x}, \overline{x}\})], & \underline{x} < 0 < \overline{x}. \end{cases}$$

2.6 Hyperbolic Tangent and Cotangent

Range. — The hyperbolic tangent and cotangent functions are defined for all *real* arguments except the hyperbolic cotangent at $x = 0$. The values of $\tanh x$ lie in the interval $(-1, 1)$, whereas $|\coth x| > 1$ for all real x. For the computed value of $\coth x$, an overflow may occur near 0. For increasing $|x|$, both functions converge very fast towards their limits ± 1.

Point Arguments. — The hyperbolic tangent and cotangent functions are computed using hyperbolic sine and cosine according to

$$\tanh x = \frac{\sinh x}{\cosh x}, \qquad \coth x = \frac{\cosh x}{\sinh x}.$$

For both functions the relative rounding error is bounded by

$$\varepsilon(\tanh) = \varepsilon(\coth) = \varepsilon(\sinh) + \varepsilon(\cosh) + \varepsilon + 4.4 \max^2 \{\varepsilon(\sinh), \varepsilon(\cosh)\}. \qquad (17)$$

The algorithm for computation is obvious. However, before dividing hyperbolic sine and cosine, the possibility of an overflow has to be checked.

Real Interval Arguments. — $\tanh x$ is strictly increasing, the hyperbolic cotangent is strictly decreasing. In case of $\coth x$, 0 must not to be part of the argument interval since 0 is a pole of the cotangent. With this exception, the interval evaluation is obvious.

2.7 Square Root

Range. — The real square root function is defined for all real $x \geq 0$. The computation of this function is important for some other real and complex functions. Hence, the square root should be implemented with the *highest available* accuracy. Computing the function value using Newton iterations is a good method for this task.

Relative Error of a Single Newton Step. — Let ε_n be the relative error of an approximation y_n for the square root $y = \sqrt{x}$, i.e.

$$y_n = \sqrt{x} \cdot (1 + \varepsilon_n).$$

Using an arithmetic with ν digits and relative error bound $\varepsilon(\nu)$, performing one Newton step yields a new approximation

$$y_{n+1} := \frac{1}{2} \cdot \left(y_n + \frac{x}{y_n} \right)$$

with the new relative error bound

$$\varepsilon_{n+1} = \left(2 + \frac{(1 + \varepsilon(\nu))^2}{2 \cdot (1 + \varepsilon_n)} + \varepsilon(\nu) \right) \cdot \varepsilon(\nu) + \frac{(1 + \varepsilon(\nu))^2}{2 \cdot (1 + \varepsilon_n)} \cdot \varepsilon_n^2. \qquad (18)$$

(18) allows choosing an appropriate arithmetical accuracy for each step. On the other hand, using (18), an error bound less than $2.5\varepsilon(\nu)$ cannot be proved. If additional Newton steps are performed, an arithmetic of increased accuracy should be used.

Initial Approximation. — The accuracy of the computed approximation depends on both the initial approximation and the number of Newton steps performed. The better the initial approximation the fewer Newton iterations are necessary and vice versa. Since there is no optimal choice for general floating-point screens, no initial approximation is given here.

A bound for the relative error ε_0 of this approximation is needed to get the number of Newton steps to be performed. This error bound has to include the theoretical approximation error as well as the rounding errors on computation of this approximation.

Algorithm. — The algorithm consists of 2 parts,

1. computation of the initial approximation, and

2. performing a number of Newton steps according to (18).

Because of the simplicity of the steps to be performed, the algorithm is not specified here. Since $\sqrt{x}$ is strictly increasing, the interval evaluation is evident.

2.8   Square

The square function with real point arguments is simply computed by multiplying the argument x by itself. The rounding error of this operation is ε.

In case of an interval argument, this method is not appropriate since multiplication of an interval argument X by itself yields an interval containing all products xy with $x, y \in X$. On the other hand, the square function is monotonically decreasing for $x < 0$ and monotonically increasing for $x > 0$ with an absolute minimum at 0. Thus, the result interval is

$$[\underline{x}, \overline{x}]^2 = \begin{cases} [\overline{x}^2, \underline{x}^2], & \overline{x} \geq 0, \\ [0, \max\{\overline{x}^2, \underline{x}^2\}], & \underline{x} < 0 < \overline{x}, \\ [\underline{x}^2, \overline{x}^2], & \overline{x} \leq 0. \end{cases}$$

Klaus Braune:

3   Complex Functions

3.1   Exponential Function, Sine and Cosine

All these functions can be handled in the same way. Differences occur only in the error bounds. In general, the functions are defined for all complex numbers.

3.1.1   Exponential Function

Range. —   With respect to the implementation, the range of the real part of the complex exponential function depends on the range of the implemented real exponential function and the range of the imaginary part on the implemented real sine and cosine functions.

Formulas and Error Bound. —   The complex exponential function is computed using the *real* functions e^x, $\sin y$ and $\cos y$ according to

$$e^z = e^{x+i \cdot y} = e^x \cdot (\cos y + i \cdot \sin y),$$
$$\operatorname{Re} e^z = e^x \cdot \cos y, \qquad \operatorname{Im} e^z = e^x \cdot \sin y,$$

yielding the error bound

$$\varepsilon(\text{cexp}) = \varepsilon(\exp) + \varepsilon(\sin) + \varepsilon + 3.1 \max{}^2 \{\varepsilon(\exp), \varepsilon(\sin)\}. \tag{19}$$

Problems occur when $e^x > 1$ and $\sin y$ or $\cos y$ are less than the smallest positive screen number, since the product may very well be a representable screen number. The problem can be solved by using an argument reduction for sine and cosine, handling all numbers of absolute value not less than $0.5 B^{em^0 - eM^0}$ like ordinary screen numbers. Numbers of smaller absolute value always cause a result underflow.

3.1.2   Sine and Cosine

Range. —   With respect to the implementation, the range of the real parts of the complex sine and cosine functions depends on the range of the implemented real sine and cosine functions, and the range of the imaginary parts depends on the implemented real hyperbolic sine and cosine functions.

Formulas and Error Bounds. —   The complex sine and cosine functions can be computed using the *real* functions $\sin x, \cos x, \sinh y$ and $\cosh y$ according to

$$\sin z = \sin x \cdot \cosh y + i \cos x \cdot \sinh y,$$
$$\cos z = \cos x \cdot \cosh y - i \sin x \cdot \sinh y,$$
$$\operatorname{Re} \sin z = \sin x \cdot \cosh y, \qquad \operatorname{Im} \sin z = \cos x \cdot \sinh y,$$
$$\operatorname{Re} \cos z = \cos x \cdot \cosh y, \qquad \operatorname{Im} \cos z = -\sin x \cdot \sinh y,$$

yielding the error bound $\varepsilon(\sin r) = \varepsilon(\cos r)$ for the real and $\varepsilon(\sin i) = \varepsilon(\cos i)$ for the imaginary part:

$$\varepsilon(\sin r) = \varepsilon(\sin) + \varepsilon(\cosh) + \varepsilon + 3.1 \ \max{}^2 \{\varepsilon(\sin), \varepsilon(\cosh)\}, \qquad (20)$$

$$\varepsilon(\sin i) = \varepsilon(\sin) + \varepsilon(\sinh) + \varepsilon + 3.1 \ \max{}^2 \{\varepsilon(\sin), \varepsilon(\sinh)\}. \qquad (21)$$

If underflow occurs for $\sin x$ or $\cos x$ and the absolute value of the corresponding factor is greater than 1, the same problems arise as with the exponential function. The problem can be fixed in the same way.

3.1.3   Complex Interval Arguments

Real and imaginary part of the complex functions e^z, $\sin z$ and $\cos z$ are representable as the product of a standard function depending only on the real part and another standard function depending only on the imaginary part of the argument. Hence, the interval functions can be computed by evaluating the appropriate *real interval* standard functions and by *interval multiplication* of the result intervals. No additional errors occur for this case with respect to the point function. An algorithm for the multiplication of real intervals can be found in [2].

3.2   Tangent and Cotangent

Range. —   Both functions are defined for all complex numbers except the poles, in case of the tangent located at $(2k + 1)\pi/2$ and in case of the cotangent at $k\pi$. These poles, except the pole of the cotangent at 0, can never be met by any number with a finite number of mantissa digits. On the other hand, floating-point numbers may be so close to poles that a result overflow occurs.

With respect to the implementation the range of the real part is restricted to the range of the real functions $\sin x$ and $\cos x$, and the imaginary part of the argument is bound to the range of the real functions $\sinh y$ and $\cosh y$.

3.2.1   Point Arguments

Formulas and Error Bounds. —   The complex tangent and cotangent functions can be computed using the *real* functions $\sin x$, $\cos x$, $\sinh y$ and $\cosh y$ according to

$$\tan z \quad = \quad \frac{\sin x \cdot \cos x}{\cos^2 x + \sinh^2 y} + i \cdot \frac{\sinh y \cdot \cosh y}{\cos^2 x + \sinh^2 y}, \qquad (22)$$

$$\cot z \quad = \quad \frac{\sin x \cdot \cos x}{\sin^2 x + \sinh^2 y} - i \cdot \frac{\sinh y \cdot \cosh y}{\sin^2 x + \sinh^2 y}. \qquad (23)$$

In case of point arguments, real and imaginary part of the cotangent can be computed by simply interchanging the trigonometric functions $\sin x$ and $\cos x$ and inverting the sign of the computed imaginary part. The required real functions have already been discussed.

Using the abbreviation

$$\varepsilon_M = \max \{|\varepsilon(\sin)|, |\varepsilon(\sinh)|, |\varepsilon(\cosh)|, |\varepsilon|\},$$

the errors of the real and imaginary part cannot exceed

$$\varepsilon(\text{ctanr}) = 4\varepsilon(\sin) + 2\varepsilon(\sinh) + 4\varepsilon + 84\varepsilon_M^2, \tag{24}$$

$$\varepsilon(\text{ctani}) = 2\varepsilon(\sin) + 3\varepsilon(\sinh) + \varepsilon(\cosh) + 4\varepsilon + 84\varepsilon_M^2. \tag{25}$$

Obviously, the error bounds of real and imaginary part are different. In general, the bigger one can be used as error bound for both parts. For the cotangent, the same error bounds $\varepsilon(\text{ccotr}) = \varepsilon(\text{ctanr})$ and $\varepsilon(\text{ccoti}) = \varepsilon(\text{ctani})$ are valid.

Problems may arise in case of an underflow of one of the values $\sin x$ or $\cos x$, since the function value itself may very well be a representable screen number.

In [3] it is shown that an underflow is recoverable only if

$$|\cos x| < \frac{1}{2} \cdot B^{em^0} \cdot \sinh^2 y.$$

A result underflow occurs without doubt if

$$|\cos x| \le 0.5 \cdot B^{3em^0}$$

With respect to cotangent, the same holds for $\sin x$.

A simple solution of this problem is to give an error message in this very improbable case. Otherwise, reduced arguments within the underflow range must be handled like ordinary screen numbers if the result may be a screen number.

Algorithm. — The complex functions tangent and cotangent can be easily computed using (22) and (23), respectively. The real functions $\sin x$, $\cos x$, $\sinh y$ and $\cosh y$ have been specified above. In case of $\cos x$ for the tangent and $\sin x$ for the cotangent, an underflow has to be handled as discussed above either by an underflow error message or by working with underflow numbers during computation. On the other hand, for $\sin x$ in case of the tangent and $\cos x$ in case of the cotangent no special underflow handling is needed.

3.2.2 Complex Interval Arguments

Special Problems. — In case of tangent and cotangent the evaluation for complex interval arguments is complicated due to

1. real and imaginary part are not monotonic but periodic with period π;

2. the points where a maximal or minimal value is attained are not representable by screen numbers and

3. can be computed only approximately.

Nevertheless, the function value can be computed with the required accuracy as proved in [3].

Obviously, argument intervals must not contain a point $z = (2k+1)\pi/2$ in case of the tangent and $k\pi$ in case of the cotangent, or an overflow error will occur.

Extreme Values. — Real and imaginary part attain their extreme values on the boundary of the argument interval, i.e. on parallels to the real or imaginary axis. Thus, it is sufficient to look for extrema on parallels to one of the axes. In the following, only the tangent is treated. For interval arguments, the appropriate method to compute the cotangent is to make use of the identity $\cot z = \tan(z - \frac{\pi}{2})$.

Real Part. — Extreme values of the real part with respect to varying x are attained at points where
$$\tan^2 x = \coth^2 y.$$
On this curve the function value is

$$\text{Re}\,\tan z \,\Big|\, {\tan^2 x = \coth^2 y} \;=\; \frac{\text{sgn}(\tan x)}{|\sinh 2y|} =: \text{sgn}(\tan x)\cdot u^*(y), \qquad (26)$$

which is a maximum if $\tan x > 0$ and a minimum if $\tan x < 0$.

For varying imaginary part the extreme values are attained on the real axis $y = 0$. The function value is obviously $\tan x$, which is a maximum in case $\tan x > 0$ and a minimum in case $\tan x < 0$.

Bounds for the Real Part. — Consideration of the monotonicity behaviour in connection with the curves just treated leads to the points where the bounds of the real part are attained. Figures showing the monotonicity behaviour are given in [3] and [4, Figure 2, page 97]. The points can be found in tables 2 – 4 together with the corresponding conditions.

The real part is even with respect to $y = \text{Im}\,z$. Hence, the discussion can be restricted to the cases $\text{Im}\,z > 0$ and $0 \in \text{Im}\,z$. In case $\text{Im}\,z < 0$ the table for

$\underline{j}$	$\overline{j}$	Conditions	Minimum
0	0	$|\tan \underline{x}| \geq \coth \overline{y}$	$u(\overline{x}, \overline{y})$
		$|\tan \overline{x}| \leq \coth \overline{y}$	$u(\underline{x}, \overline{y})$
		otherwise	$u(\underline{x}|\overline{x}, \overline{y})$
	1	$|\tan \overline{x}| > \coth y$	$u(\overline{x}, y)$
		otherwise	$-u^*(y)$
	≥ 2		$-u^*(y)$
1	1	$|\tan \underline{x}| < \coth y$	$u(\underline{x}, y)$
		$|\tan \overline{x}| > \coth y$	$u(\overline{x}, y)$
		otherwise	$-u^*(y)$
	2	$|\tan \underline{x}| < \coth y$	$u(\underline{x}, y)$
		otherwise	$-u^*(y)$
	3	$|\tan \underline{x}| < \coth y \wedge |\tan \overline{x}| > \coth \overline{y}$	$u(\underline{x}|\overline{x}, y)$
		otherwise	$-u^*(y)$
	≥ 4		$-u^*(y)$

Table 2: Lower bound of $\text{Re}\,\tan z$ in case $\text{Im}\,z > 0$.

$\underline{j}$	$\bar{j}$	Conditions	Maximum
0	0	$|\tan \bar{x}| < \coth y$	$u(\bar{x}, y)$
		$|\tan \underline{x}| > \coth y$	$u(\underline{x}, y)$
		otherwise	$u^*(y)$
	1	$|\tan \underline{x}| > \coth y$	$u(\underline{x}, y)$
		otherwise	$u^*(y)$
	2	$|\tan \underline{x}| > \coth y \wedge |\tan \bar{x}| < \coth y$	$u(\underline{x}|\bar{x}, y)$
		otherwise	$u^*(y)$
	≥ 3		$u^*(y)$
1	1	$|\tan \underline{x}| \leq \coth \bar{y}$	$u(\bar{x}, \bar{y})$
		$|\tan \bar{x}| \geq \coth \bar{y}$	$u(\underline{x}, \bar{y})$
		otherwise	$u(\underline{x}|\bar{x}, \bar{y})$
	2	$|\tan \bar{x}| < \coth y$	$u(\bar{x}, y)$
		otherwise	$u^*(y)$
	≥ 3		$u^*(y)$

Table 3: Upper bound of Re tan z in case Im z > 0.

Im $z > 0$ can be used with y and $\bar{y}$ interchanged. The real part tends to 0 for y tending to ∞. In particular, Re tan z always lies between 0 and tan x.

The points where the real part of the tangent attains its bounds are determined by the following tables. The cases are separated mainly according to the entier parts $\underline{j}$ and $\bar{j}$ at the lower and upper bound of the real part of the argument. The relation $\underline{j} \leq \bar{j}$ is assumed. If this relation does not hold, $\bar{j}$ is increased by 4. Sometimes there are two points where the corresponding bound may be reached. In this case the real or imaginary parts of both points are specified. In case of a lower bound the smaller and in case of an upper bound the greater value has to be chosen.

In case $0 \in \text{Im} z$ the real part must not contain one of the values $(2k+1)\pi/2$, or an overflow error will occur due to a pole within the argument. In particular, $\bar{x} - \underline{x} < \pi$ must hold. In case $\tan \underline{x} \leq 0$ this value is the minimum of the real part, and in case $\tan \bar{x} \geq 0$ this value is the maximum. In all other cases the extrema are attained for the imaginary part

$$y^* := \max\{-\underline{y}, \bar{y}\}.$$

Because of the poles between $j = 0$ and $j = 1$, the relation $\bar{j} \geq \underline{j} \geq 1$ is assumed. $\bar{j} \geq 3$ always leads to an error because of the pole between $j = 2$ and $j = 3$.

Imaginary Part. — For the imaginary part in case of varying x the extrema are located on the curves $x = k\pi/2$. Obviously, the function value on the curves is

$$\text{Im} \tan z \Big|_{x = k\pi/2} = \begin{cases} \tanh y, & k = 2n, \\ \coth y, & k = 2n+1. \end{cases} \tag{27}$$

For $y > 0$ the value tanh y is a minimum, whereas coth y is a maximum. On the other hand, for $y < 0$ the value coth y is a minimum and tanh y a maximum.

$\underline{j}$	$\bar{j}$	Conditions	Minimum	Maximum
1	1	$|\tan \underline{x}| \leq \coth y^*$	$\tan \underline{x}$	$u(\bar{x}, y^*)$
		$|\tan \bar{x}| \geq \coth y^*$		$u(\underline{x}, y^*)$
		otherwise		$u(\underline{x}|\bar{x}, y^*)$
	2			$\tan \bar{x}$
2	2	$|\tan \underline{x}| \geq \coth y^*$	$u(\bar{x}, y^*)$	
		$|\tan \bar{x}| \leq \coth y^*$	$u(\underline{x}, y^*)$	
		otherwise	$u(\underline{x}|\bar{x}, y^*)$	

Table 4: Bounds of Re $\tan z$ in case $0 \in \operatorname{Im} z$.

Thus, in each case the imaginary part for varying x and fixed y is situated between the values $\tanh y$ and $\coth y$.

For varying imaginary part of the argument the extreme values are attained on the curve $\tan^2 x = \coth^2 y$ already known from the discussion of the real part. The function value is

$$\left. \operatorname{Im} \tan z \right|_{\tan^2 x = \coth^2 y} = \frac{\operatorname{sgn} y}{|\sin 2x|} =: \operatorname{sgn} y \cdot v^*(x), \tag{28}$$

which is a maximum in case $y > 0$ and a minimum in case $y < 0$.

Bounds for the Imaginary Part. — The points where the bounds of the imaginary part are reached are found by considering the monotonicity behaviour in connection with the curves just treated. Figures showing the monotonicity behaviour can be found in [3] and [4, Figure 3, page 100]. The points are specified by tables 5 – 8 together with the corresponding conditions.

The imaginary part is odd with respect to $\operatorname{Im} z$. Hence, only the cases $\operatorname{Im} z > 0$ and $0 \in \operatorname{Im} z$ have to be treated. In case $\operatorname{Im} z < 0$ the table for $\operatorname{Im} z > 0$ can be used with y and $\bar{y}$ interchanged and the sign of the result changed. The imaginary part tends to ± 1 for y tending to $\pm\infty$. In particular, $\operatorname{Im} \tan z$ always lies between $\tanh y$ and $\coth y$.

The points where the imaginary part of the tangent reaches its bounds are determined by the following tables. The same assumptions as for the real part are made.

In case $0 \in \operatorname{Im} z$ the conditions $(2k + 1)\pi/2 \notin \operatorname{Re} z$ and $\bar{x} - \underline{x} < \pi$ must hold. Since the sign of the imaginary part is equal to the sign of $\operatorname{Im} z$ for each point, the minimum is always attained for $\operatorname{Im} z \leq 0$ and the maximum for $\operatorname{Im} z \geq 0$.

Error Bounds. — Since special formulas are used in case of interval arguments and additional errors occur due to problems with finding extreme function values, the error bounds for point arguments are not valid for intervals. However, in [3] the relative error bounds

$$\varepsilon(\text{citanr}) = 8\varepsilon(\sin) + 4\varepsilon(\sinh) + 2\varepsilon(\cosh) + 8\varepsilon + 555\varepsilon_M^2, \tag{29}$$

$$\varepsilon(\text{citani}) = 6\varepsilon(\sin) + 5\varepsilon(\sinh) + 3\varepsilon(\cosh) + 8\varepsilon + 555\varepsilon_M^2. \tag{30}$$

are proved to hold for complex interval arguments.

$\underline{j}$	$\overline{j}$	Conditions		Minimum
0	0	$\lvert\tan \underline{x}\rvert \le \coth \overline{y}$		$v(\underline{x}, \underline{y})$
		$\lvert\tan \underline{x}\rvert \ge \coth \underline{y}$		$v(\underline{x}, \overline{y})$
		otherwise		$v(\underline{x}, \underline{y}\vert\overline{y})$
	1	$\lvert\tan \underline{x}\rvert \le \coth \overline{y}$ $\vee \lvert\tan \overline{x}\rvert \le \coth \overline{y}$	$\lvert\tan \underline{x}\rvert \le \lvert\tan \overline{x}\rvert$	$v(\underline{x}, \underline{y})$
			otherwise	$v(\overline{x}, \underline{y})$
		$\lvert\tan \underline{x}\rvert \ge \coth \underline{y}$ $\wedge \lvert\tan \overline{x}\rvert \ge \coth \underline{y}$	$\lvert\tan \underline{x}\rvert \le \lvert\tan \overline{x}\rvert$	$v(\underline{x}, \overline{y})$
			otherwise	$v(\overline{x}, \overline{y})$
		otherwise	$\lvert\tan \underline{x}\rvert \le \lvert\tan \overline{x}\rvert$	$v(\underline{x}, \underline{y}\vert\overline{y})$
			otherwise	$v(\overline{x}, \underline{y}\vert\overline{y})$
	≥ 2			$\tanh y$
1	1	$\lvert\tan \overline{x}\rvert \le \coth \overline{y}$		$v(\overline{x}, \underline{y})$
		$\lvert\tan \overline{x}\rvert \ge \coth \underline{y}$		$v(\overline{x}, \overline{y})$
		otherwise		$v(\overline{x}, \underline{y}\vert\overline{y})$
	≥ 2			$\tanh \underline{y}$

Table 5: Lower bound of Im $\tan z$ in case Im $z > 0$.

$\underline{j}$	$\overline{j}$	Conditions		Maximum
0	0	$\lvert\tan \overline{x}\rvert < \coth \overline{y}$		$v(\overline{x}, \overline{y})$
		$\lvert\tan \overline{x}\rvert > \coth \underline{y}$		$v(\overline{x}, \underline{y})$
		otherwise		$v^*(\overline{x})$
	≥ 1			$\coth y$
1	1	$\lvert\tan \underline{x}\rvert < \coth \overline{y}$		$v(\underline{x}, \overline{y})$
		$\lvert\tan \underline{x}\rvert > \coth \underline{y}$		$v(\underline{x}, \underline{y})$
		otherwise		$v^*(\underline{x})$
	2	$\lvert\tan \underline{x}\rvert > \coth \underline{y}$ $\vee \lvert\tan \overline{x}\rvert > \coth \underline{y}$	$\lvert\tan \underline{x}\rvert \ge \lvert\tan \overline{x}\rvert$	$v(\underline{x}, \underline{y})$
			otherwise	$v(\overline{x}, \underline{y})$
		$\lvert\tan \underline{x}\rvert < \coth \overline{y}$ $\wedge \lvert\tan \overline{x}\rvert < \coth \overline{y}$	$\lvert\tan \underline{x}\rvert \ge \lvert\tan \overline{x}\rvert$	$v(\underline{x}, \overline{y})$
			otherwise	$v(\overline{x}, \overline{y})$
		otherwise	$\lvert\tan \underline{x}\rvert \ge \lvert\tan \overline{x}\rvert$	$v^*(\underline{x})$
			otherwise	$v^*(\overline{x})$
	≥ 3			$\coth y$

Table 6: Upper bound of Im $\tan z$ in case Im $z > 0$.

Algorithm. — The algorithm for interval evaluation is based on tables 2 – 8. The tables determine which function is to be evaluated at what point to find bounds for the real and for the imaginary part.

1. Evaluate the real functions $\sinh y$, $\cosh y$ and $\tanh y := \sinh y / \cosh y$, $\coth y := \cosh y / \sinh y$ at $\underline{y}$ and $\overline{y}$.

2. Compute $\cos x$ with underflow exponent $U = 0.5 B^{3em^0}$ at $\underline{x}$ and $\overline{x}$. Compute $\sin x$ with underflow exponent $U = 0.5 B^{em^0}$ at $\underline{x}$ and $\overline{x}$ together with the entier parts $\underline{j}$ and $\overline{j}$.

$\underline{j}$	$\bar{j}$	Conditions			Minimum
1	1	$|\tan \underline{x}| < -\coth y$			$v(\underline{x}, y)$
		otherwise			$-v^{\bullet}(\underline{x})$
	2	$|\tan \underline{x}| < |\tan \bar{x}|$	$|\tan \bar{x}| < -\coth y$		$v(\bar{x}, y)$
			otherwise		$-v^{\bullet}(\bar{x})$
		otherwise	$|\tan \underline{x}| < -\coth y$		$v(\underline{x}, \bar{y})$
			otherwise		$-v^{\bullet}(\underline{x})$
2	2	$|\tan \bar{x}| < -\coth y$			$v(\underline{x}, \bar{y})$
		otherwise			$-v^{\bullet}(\bar{x})$

Table 7: Lower bound of Im tan z in case $0 \in \mathrm{Im}\, z$.

$\underline{j}$	$\bar{j}$	Conditions			Maximum
1	1	$|\tan \underline{x}| < \coth \bar{y}$			$v(\underline{x}, \bar{y})$
		otherwise			$v^{\bullet}(\underline{x})$
	2	$|\tan \underline{x}| < |\tan \bar{x}|$	$|\tan \bar{x}| < \coth \bar{y}$		$v(\bar{x}, \bar{y})$
			otherwise		$v^{\bullet}(\bar{x})$
		otherwise	$|\tan \underline{x}| < \coth \bar{y}$		$v(\underline{x}, \bar{y})$
			otherwise		$v^{\bullet}(\underline{x})$
2	2	$|\tan \bar{x}| < \coth \bar{y}$			$v(\bar{x}, \bar{y})$
		otherwise			$v^{\bullet}(\bar{x})$

Table 8: Upper bound of Im tan z in case $0 \in \mathrm{Im}\, z$.

3. In case $0 \in \mathrm{Im}\, z$ check and handle the error conditions $\mathrm{Re}\,\bar{z} - \mathrm{Re}\,\underline{z} > \pi$ or $\bar{j} - \underline{j} > 4$.

4. In case $\bar{j} < \underline{j}$, increase $\bar{j}$ by 8.

5. Replace $\bar{j}$ by $(\underline{j} \bmod 2) + \bar{j} - \underline{j}$ and $\underline{j}$ by $\underline{j} \bmod 2$.

6. Compute $\tan \underline{x}$ and $\tan \bar{x}$ by dividing the previously computed values of sine and cosine at these points.

7. Determine from tables 2 – 8 the points for computing the bounds of real and imaginary part and compute the bounds according to (22) and (26) – (28).

If the cotangent is to be computed, first compute $\tan(z - \frac{\pi}{2})$ using $\sin(x - \frac{\pi}{2}) = -\cos x$ and $\cos(x - \frac{\pi}{2}) = \sin x$. The entier part of $x - \frac{\pi}{2}$ is $j - 1$. Finally, change the sign of the real and the imaginary part by inverting the sign of each bound and by interchanging lower and upper bounds.

3.3   Hyperbolic Functions

The complex hyperbolic functions are strongly related to the trigonometric functions. Hyperbolic sine and cosine as well as hyperbolic tangent and cotangent can be computed using the corresponding trigonometric functions. The only additional operations to be performed are changing signs and interchanging real and imaginary part which cause no additional errors. The specific formulas are:

$$\sinh z = \text{Im } \sin(\text{Im } z + i \cdot \text{Re } z) + i \cdot \text{Re } \sin(\text{Im } z + i \cdot \text{Re } z),$$
$$\cosh z = \text{Re } \cos(\text{Im } z + i \cdot \text{Re } z) - i \cdot \text{Im } \cos(\text{Im } z + i \cdot \text{Re } z),$$
$$\tanh z = \text{Im } \tan(\text{Im } z + i \cdot \text{Re } z) + i \cdot \text{Re } \tan(\text{Im } z + i \cdot \text{Re } z),$$
$$\coth z = -\text{Im } \cot(-\text{Im } z + i \cdot \text{Re } z) + i \cdot \text{Re } \cot(-\text{Im } z + i \cdot \text{Re } z).$$

These formulas allow the computation of the hyperbolic functions using the algorithms for the corresponding trigonometric function.

3.4 Square Root

Range. — The complex square root is defined for all complex numbers. In general, the complex plane is mapped onto the right half-plane. The branch cut along the negative real axis is mapped onto the positive imaginary axis.

Formulas and Error Bound. — The complex square root function can be computed using the *real* square root according to

$$\sqrt{z} = \frac{1}{\sqrt{2}} \cdot \left(\sqrt{|z| + x} + i \cdot \frac{y}{\sqrt{|z| + x}} \right),$$

$$\text{Re } \sqrt{z} = \frac{1}{\sqrt{2}} \cdot \sqrt{|z| + x}, \qquad \text{Im } \sqrt{z} = \frac{1}{\sqrt{2}} \cdot \frac{y}{\sqrt{|z| + x}}. \tag{31}$$

If in case $x < 0$ leading digits of the term $|z| + x$ are cancelled, use the identity

$$\sqrt{|z| - |x|} = \frac{|y|}{\sqrt{|z| + |x|}},$$

i.e the formulas of real and imaginary part are interchanged. The algorithm for the computation of the complex square root is obvious.

The relative error when computing according to these formulas is bounded by

$$\varepsilon(\text{csqrt}) = 1.51\varepsilon(\text{sqrt}) + 4.1\varepsilon. \tag{32}$$

This error bound includes the error due to the approximation of the constant $1/\sqrt{2}$ with error bound ε.

Complex Interval Arguments. — In case of complex interval arguments, problems arise if the negative real axis has points in common with the argument interval and the argument does not contain 0. In this case, according to the result range specified for the point function, the points above and on the real axis would be mapped into the upper right quadrant and the points below the real axis into the lower right quadrant. Since 0 is not part of the argument

Conditions	LB(Re)	UB(Re)	LB(Im)	UB(Im)
$\underline{y} \geq 0$	$\underline{x}, \underline{y}$	$\overline{x}, \overline{y}$	$\overline{x}, \underline{y}$	$\underline{x}, \overline{y}$
$\overline{y} \leq 0$	$\underline{x}, \overline{y}$	$\overline{x}, \underline{y}$	$\underline{x}, \underline{y}$	$\overline{x}, \overline{y}$
$\underline{y} < 0 < \overline{y} \quad \underline{x} \geq 0$	$\underline{x}, 0$	$\overline{x}, My$ [1]	$\underline{x}, \underline{y}$	$\underline{x}, \overline{y}$
$\overline{x} \leq 0$	$\overline{x}, \underline{y}$ [2]	$\overline{x}, \overline{y}$	$\overline{x}, 0$	$\underline{x}, My$ [1,2]
$\underline{x} < 0 < \overline{x}$	$0, 0$	$\overline{x}, My$ [1]	$\underline{x}, \underline{y}$	$\underline{x}, \overline{y}$

[1] $My = \underline{y}$, if $|\underline{y}| > |\overline{y}|$, $My = \overline{y}$ otherwise.

[2] Compute $\underline{y}$ according to (33).

Table 9: Points for interval evaluation of complex square root.

interval, the two parts are *not connected*. Thus, a result interval enclosing both parts overestimates the true result, sometimes even in an extreme way.

To avoid overestimation in this special case, the image of the argument interval is chosen to be part of the *upper plane*, i.e. points with both real and imaginary part negative are mapped onto the upper left quadrant of the complex plane yielding a connected image of the argument interval. If the boundary of an argument interval has points in common with the negative real axis but not with the upper half plane, the negative real axis is mapped onto the negative imaginary axis, i.e. is treated as part of the lower half plane.

For such intervals, inclusion monotonicity is lost, but can be recovered by computation of the square root of the part on and above the real axis and that below the real axis separately and taking their convex hull.

The formulas for mapping the lower left quadrant into the upper left quadrant are

$$\sqrt{z} = \frac{1}{\sqrt{2}} \cdot \left(\frac{y}{\sqrt{|z| + |x|}} + i \cdot \sqrt{|z| + |x|} \right),$$

$$\mathrm{Re}\,\sqrt{z} = \frac{1}{\sqrt{2}} \cdot \frac{y}{\sqrt{|z| + |x|}}, \qquad \mathrm{Im}\,\sqrt{z} = \frac{1}{\sqrt{2}} \cdot \sqrt{|z| + |x|}, \qquad (33)$$

obviously with the same error bound as the formulas (31).

Moreover, both the real and the imaginary part are not monotonic. Table 9 gives the points where the lower and upper bound $\dot{L}B(\mathrm{Re})$ and $UB(\mathrm{Re})$ of the real part and $LB(\mathrm{Im})$ and $UB(\mathrm{Im})$ of the imaginary part are attained.

The algorithm for computation of the square root of an interval can easily be implemented using the formulas (31) and (33) according to table 9.

3.5 Square

Range. — The complex square is defined for all complex numbers. However, for large arguments an overflow may occur.

Conditions		LB(Re)	UB(Re)	LB(Im)	UB(Im)
$\underline{y} \geq 0$	$\underline{x} \geq 0$	$\underline{x}, \overline{y}$	$\overline{x}, \underline{y}$	$\underline{x}, \underline{y}$	$\overline{x}, \overline{y}$
	$\overline{x} \leq 0$	$\overline{x}, \overline{y}$	$\underline{x}, \underline{y}$	$\underline{x}, \overline{y}$	$\overline{x}, \underline{y}$
	$\underline{x} < 0 < \overline{x}$	$0, \overline{y}$	$Mx, \underline{y}$ [1]	$\underline{x}, \overline{y}$	$\overline{x}, \overline{y}$
$\overline{y} \leq 0$	$\underline{x} \geq 0$	$\underline{x}, \underline{y}$	$\overline{x}, \overline{y}$	$\overline{x}, \underline{y}$	$\underline{x}, \overline{y}$
	$\overline{x} \leq 0$	$\overline{x}, \underline{y}$	$\underline{x}, \overline{y}$	$\overline{x}, \overline{y}$	$\underline{x}, \underline{y}$
	$\underline{x} < 0 < \overline{x}$	$0, \underline{y}$	$Mx, \overline{y}$ [1]	$\overline{x}, \underline{y}$	$\underline{x}, \underline{y}$
$\underline{y} < 0 < \overline{y}$	$\underline{x} \geq 0$	$\underline{x}, My$ [1]	$\overline{x}, 0$	$\overline{x}, \underline{y}$	$\overline{x}, \overline{y}$
	$\overline{x} \leq 0$	$\overline{x}, My$ [1]	$\underline{x}, 0$	$\underline{x}, \overline{y}$	$\underline{x}, \underline{y}$
	$\underline{x} < 0 < \overline{x}$	$0, My$ [1]	$Mx, 0$ [1]	min [2]	max [2]

[1] $Mx = \max(|\underline{x}|, |\overline{x}|), My = \max(|\underline{y}|, |\overline{y}|)$.

[2] $\min = \min(\underline{x} \cdot \overline{y}, \overline{x} \cdot \underline{y}), \max = \max(\underline{x} \cdot \underline{y}, \overline{x} \cdot \overline{y})$.

Table 10: Points for interval evaluation of z^2.

Formulas and Error Bounds. — The complex square function can be computed using *real* arithmetic according to

$$z^2 = (x - y) \cdot (x + y) + 2ixy, \tag{34}$$

with error $\varepsilon(\text{csqrr})$ for the real and $\varepsilon(\text{csqri})$ for the imaginary part bounded by

$$\varepsilon(\text{csqrr}) = 3.1\,\varepsilon, \qquad \varepsilon(\text{csqri}) = 2.1\,\varepsilon. \tag{35}$$

The algorithm for point arguments is obvious.

Complex Interval Arguments. — The imaginary part can be computed by interval multiplication of the real and the imaginary part of the argument and multiplication of both bounds by 2. However, the real part is neither monotonic nor can it be represented as a product of factors depending either on the real or the imaginary part. Table 10 specifies which points to choose for computation of the lower and upper bound of the real part. The corresponding points for the computation of the bounds of the imaginary part are included for the computation of the imaginary part without using interval multiplication.

The algorithm for computing the square of an interval can easily be implemented using the formulas (35) according to table 10.

4 Number of Guard Digits

Depending on the base the preceding error estimates for the functions determine a sufficient number of mantissa digits of the screen $\mathscr{S}$ to return — with few exceptions — in case of point interval arguments 2 neighboring floating-point numbers as guaranteed bounds for the exact result. The following tables are based on a total relative error never exceeding $\varepsilon(\ell^0)/100 = B^{1-\ell^0}/100$.

Function	used	$\varepsilon(app)$	Conditions	k_{min}
e^x	—	$0.5\varepsilon(\ell - 1)$	$B^k \geq \frac{10.21\|x\|+1.021}{\ln B} B^{n+\ell}$	$1 + \frac{5.71}{\ln B}$
$\sin x,$ $\cos x$	—	$0.6\varepsilon(\ell - 1)$	—	$1 + \frac{7.01}{\ln B}$
$\tan x,$ $\cot x$	$\sin x,$ $\cos x$	—	—	$1 + \frac{7.81}{\ln B}$
$\sinh x$	e^t	$3.1\varepsilon(\ell - 1)$	—	$1 + \frac{6.06}{\ln B}$
$\cosh x$	e^t	—	—	$1 + \frac{6.05}{\ln B}$
$\tanh x,$ $\coth x$	$\sinh x,$ $\cosh x, e^t$	—	—	$1 + \frac{6.80}{\ln B}$
$\sqrt{x}$	—	iterative	iterated until $\varepsilon(sqrt) = 6\varepsilon$	$\frac{6.40}{\ln B}$
x^2	—	—	—	$\frac{4.61}{\ln B}$

Table 11: Minimal number of guard digits for real functions.

Function	used	Conditions	k_{min}
e^z	$e^x, \sin y, \cos y$	—	$1 + \frac{7.28}{\ln B}$
$\sin z, \cos z,$ $\sinh z, \cosh z$	$\sin x, \cos x, e^t,$ $\sinh y, \cosh y$	—	$1 + \frac{7.37}{\ln B}$
$\tan z, \cot z,$ $\tanh z, \coth z$	$\sin x, \cos x, e^t,$ $\sinh y, \cosh y$	—	$1 + \frac{8.62}{\ln B}$ [1] $1 + \frac{9.43}{\ln B}$ [2]
$\sqrt{z}$	$\sqrt{t}$	$\varepsilon(sqrt) = 6\varepsilon$	$\frac{7.19}{\ln B}$
z^2	—	—	$\frac{5.74}{\ln B}$

[1] For complex point functions.

[2] For complex interval functions.

Table 12: Minimal number of guard digits for complex functions.

The following conclusions hold if *all errors* are *less than* 10^{-4}, including the bounds for the error of the function itself. Moreover, approximation and reduction errors are assumed to be approximately of the same order as the rounding errors. Special requirements are specified with the table.

The last column of the table gives a lower bound for the number

$$k := \ell - \ell^0$$

of necessary additional mantissa digits for this case. Another column shows the functions used to compute the function value. These (real) functions have to be computed with the specified accuracy, too. The required approximation accuracy in case of the real functions is specified depending on ε.

References

[1] IBM *High-Accuracy Arithmetic Subroutine Library (ACRITH). Program Description and User's Guide.* SC33-6164-02, 3rd Edition, April 1986.

[2] Alefeld, G., Herzberger J.: *An Introduction to Interval Computations.* Academic Press, New York, 1983.

[3] Braune, K.: *Hochgenaue Standardfunktionen für reelle und komplexe Punkte und Intervalle in beliebigen Gleitpunktrastern — Exponentialfunktion, Trigonometrische und Hyperbolische Funktionen, Wurzel und Quadrat.* Dissertation, Universität Karlsruhe, 1987.

[4] Braune, K., Krämer, W.: High-accuracy standard functions for real and complex intervals. In Kaucher, E., Kulisch, U., Ullrich, Ch. (eds.): *Computerarithmetic, Scientific Computation and Programming Languages,* pages 81–114. B. G. Teubner Verlag, Stuttgart, 1987.

[5] Brent, R.P.: *MP User's Guide.* Technical Report TR-CS-81-08, Department of Computer Science, Australian National University, Canberra, 1981.

[6] Krämer, W.: *Inverse Standard Functions for Real and Complex Point and Interval Arguments with Dynamic Accuracy.* This Volume.

[7] Krämer, W.: *Inverse Standardfunktionen für reelle und komplexe Intervallargumente mit a priori Fehlerabschätzungen für beliebige Datenformate.* Dissertation, Universität Karlsruhe, 1987.

Dr. Klaus D. Braune
Institut für Angewandte Mathematik
Universität Karlsruhe
Englerstrasse 2
D-7500 Karlsruhe
Federal Republic of Germany

Computing, Suppl. 6, 185–212 (1988)

Inverse Standard Functions for Real and Complex Point and Interval Arguments with Dynamic Accuracy *

Walter Krämer, Karlsruhe

Abstract — Zusammenfassung

Inverse standard functions for real and complex point and interval arguments with dynamic accuracy.
Algorithms to compute inverse standard functions to arbitrary accuracy with safe error bounds
are given. Not only approximation errors but also all possible rounding errors are considered. The
desired accuracy of the function as well as the base of the number system used are parameters
of the error formula. For implementation it is only assumed that the four elementary arithmetic
operations are performed with a certain number of correct digits, which is given by the error formula
and little higher than the desired number of correct digits of the function value. The interval routines
are constructed out of the point routines considering the monotonicity behaviour of the functions.
The ambiguity of the complex functions is briefly discussed (for a detailed discussion see [4] and
[5]).

**Inverse Standardfunktionen für reelle und komplexe Punkt- und Intervallargumente mit dynamischer
Genauigkeit.** Es werden Algorithmen zur Berechnung von inversen Standardfunktionen auf beliebige
Länge mit bewiesenen Fehlerschranken besprochen. Dabei werden nicht nur Approximationsfehler
sondern auch alle auftretenden Rundungsfehler a priori mitberücksichtigt. Die gewünschte Genauigkeit der Funktion sowie die Basis des verwendeten Zahlsystems gehen als Parameter in die Fehlerfor-
meln ein. Zur Implementierung wird nur vorausgesetzt, daß die vier Grundrechenarten auf eine
durch die Fehlerformel bestimmte Stellenzahl, welche nur leicht über der für das Ergebnis der Stan-
dardfunktion gewünschten Stellenzahl liegt, verfügbar sind. Die Intervallroutinen werden unter Zuhil-
fenahme von Monotoniebetrachtungen aus den Punktroutinen aufgebaut. Schwierigkeiten, die sich
aus der Vieldeutigkeit der komplexen Funktionen ergeben, werden kurz gestreift; näheres siehe [4]
und [5].

1. Introduction

The standard functions discussed in this paper and in Braune [3] have been
implemented for 2 specific floating-point data formats as part of the IBM program product ACRITH [1]. The mathematical problems occurring in this implementation have been presented in [4]. The definitions of the terms used in
this paper like "real and complex interval function', 'branch cut' and 'many-
valued functions' can be found in [4] together with illustrating figures.

These methods are extended in [2] and [5] to arbitrary floating-point data
formats including rigorous error bounds.

* Diese Arbeit wurde mit Mitteln des Ministeriums für Wirtschaft, Mittelstand und Technologie
Baden-Württemberg gefördert.

In this paper the *implementation* of the standard functions for arbitrary data formats is treated. Bounds for the error when computing the function value using the specified algorithm are given. All given formulas are proved in [2] and [5].

1.1 Motivation

Nowadays, standard functions in electronic computers are considered to be part of computer arithmetic. Nevertheless, in most cases, the accuracy of the implemented functions is not known by the user. In general, this is true for the manufacturer himself. Only random tests are performed and published, giving a feeling but not a certainty of the number of correct mantissa digits in the result of a function call.

Of course, the best result would be a floating point number which is nearest to the exact result (1/2 ulp accuracy, i.e. the error is at most 1/2 unit in the last place of the mantissa). Because the considered functions are transcendental this goal is not attainable. The same is true for function values, to be rounded downwards or upwards to the next floating-point number below or above (interval arithmetic).

Nevertheless it is possible to implement all the considered routines in such a way that the result for point input data gives a number which is one of the adjacent floating-point numbers of the exact result; i.e. the accuracy is better than 1 ulp. When directed rounding is required, in the worst case one additional floating-point number lies between the rounded result and the exact function value (accuracy better than 2 ulp). This can be achieved not only for the real functions but also for the real and imaginary parts of the complex functions.

The real-valued functions

$$\log(X), X^Y, \text{abs}(Z), \arg(Z),$$
$$\arctan(X), \text{arccot}(X),$$
$$\text{arsinh}(X), \text{arcosh}(X), \text{artanh}(X), \text{arcoth}(X),$$

as well as the complex-valued functions

$$\log(Z), \arcsin(Z) \text{ and } \arctan(Z)$$

are considered in this article. Other functions like $\arctan 2(X, Y)$, $\log 10(Z)$, $\arccos(Z)$, $\text{arccot}(Z)$, $\text{arsinh}(Z)$, $\text{arcosh}(Z)$, $\text{artanh}(Z)$ and $\text{arcoth}(Z)$ may be found in [5]. Thereby, the input argument Z denotes a complex point argument or a complex interval argument. X, Y denote real point or real interval arguments. The algorithms given are widely independent of the data format, i.e. of the used base B and the used mantissa length. Therefore, it should be possible to implement the algorithms in a simple way and to have valid bounds for the maximal relative error of such a routine.

Another important possibility is to implement functions for arbitrary mantissa length. This, for example, may be necessary, when the solution of a nonlinear system has to be determined. In this case serious cancellation during computation of the approximate solution may be avoided using standard function routines for appropriate mantissa length. An interval inclusion of the solution may be verified using interval versions of the function routines.

1.2 Notation and Conventions

Implementing highly accurate functions for a fixed data format is usually done using two floating-point screens. The less accurate I/O-screen S^0 has $l^0 = l - k$ mantissa digits and the internally used more accurate approximation screen S uses l mantissa digits to the base B. Especially all screen numbers of S^0 are also elements of the approximation screen and, therefore, are representable in S without rounding error.

Of course, statements about the accuracy of functions are only valid for exactly representable input arguments. Possible conversion errors, for example when changing from decimal data to internal hexadecimal or binary representation, have to be handled separately.

- $S^0(B, l^0, em^0, eM^0)$ means the I/O screen. B is the base, $l^0 > 0$ the number of mantissa digits, em^0 the minimal and eM^0 the maximal exponent of the number system.
- $S(B, l, ., .)$ analogously characterizes the finer internal approximation screen. The exponent range is assumed to be large enough and is not further specified.
- $\varepsilon := B^{-l+1}$ denotes a bound for the relative rounding error of the operations $+$, $-$, $*$, and $/$ in the approximation screen. If the arithmetic is of maximum accuracy [6], the value of ε may be halved
- **MINREAL** denotes the smallest positive floating-point number of S^0, **MAXREAL** the largest one.
- For an operation $\cdot \in \{+, -, *, /\}$, the estimation

$$\left| \frac{(a \cdot b) - (a \odot b)}{a \cdot b} \right| \leq \varepsilon$$

holds. $\odot$ means the rounded operation. Furthermore

$$\left| \frac{c - \tilde{c}}{c} \right| \leq \varepsilon$$

is assumed for any constant c and the corresponding machine approximation $\tilde{c}$.
- $\tilde{x}$ denotes an approximation (afflicted with rounding and/or approximation error) of the associated value x. Normally $\tilde{x}$ is the machine representation of x. Analogously, $\tilde{f}$ and $\tilde{\tilde{f}}$ are used for a function $f(x)$.
- $\varepsilon(f)$ is an upper bound of the relative error when computing the function $f(x)$ on a computer, i.e.

$$\left| \frac{f(x) - \tilde{f}(x)}{f(x)} \right| \leq \varepsilon(f).$$

● Taylor polynomials which are used to approximate $f(x)$ are always computed using Horner's scheme. In most cases f is approximated in the form $f(x)$ $\approx x \cdot \sum_{n=0}^{N} a_n \cdot (x^2)^n$. The number N is called the degree of the approximation (this, in general, is not the degree of the polynomial). The error bounds are only valid for $N < \frac{1}{2}\left(\sqrt{\dfrac{2}{B \cdot \varepsilon}} - 1\right)$, which is not a serious restriction. $\varepsilon(\text{app})$ denotes an upper bound for the relative error when truncating the series after $N+1$ summands, i.e.

$$\left| \frac{f(x) - x \cdot \sum_{n=0}^{N} a_n \cdot (x^2)^n}{f(x)} \right| < \varepsilon(\textbf{app}).$$

● For almost every function, $\varepsilon(\text{app})$ is part of the final error formula. If the relative error $\varepsilon(f)$ of a specific function is to be less than B^{1-l}, the degree N of the used polynomial approximation ($|x| < \gamma < 1/8$, where x is the actual reduced argument) is to be determined in such a way, that the corresponding $\varepsilon(\text{app})$ is small enough. The following formula may be used for all quadratic approximations used in this article (worst case and, therefore, not optimal).

$$N = \min\left\{ n \in \mathbb{N}, \left| \frac{\gamma^{2n+2}}{|a_0| - |a_1| \cdot \gamma^2} \right| < \varepsilon(\text{app}) \right\}.$$

● Only floating-point screens with

$$\varepsilon = \varepsilon(m) = B^{-m+1} \quad \text{and} \quad \varepsilon \cdot B < 10^{-3}$$

are considered. Such screens are finer than a decimal screen with 4 mantissa digits.

● The requirement upon the accuracy for a complex valued function $f(z)$ is not only coupled with a relative error for $|f(z)|$. The following stronger requirement must hold:

$$\left| \frac{\text{Re}(f(z)) - \text{Re}(\tilde{f}(z))}{\text{Re}(f(z))} \right| < \varepsilon, \qquad \left| \frac{\text{Im}(f(z)) - \text{Im}(\tilde{f}(z))}{\text{Im}(f(z))} \right| < \varepsilon.$$

The definition of the term 'extremal curve' may be found in [4].

2. Real Standard Functions

2.0 Computation of Bounds for Real Standard Functions

Let $\tilde{f}(x) \in S$ be an approximation to the exact function value $f(x)$ with a relative error less than $\varepsilon(f)$. Then an upper bound $\hat{f}(x) \in S$ to the exact function value is given by:

$$\hat{f}(x) = \begin{cases} \tilde{f}(x) \circledast \triangle(1 + \varepsilon(f) + 1.1\varepsilon) & \text{for } \tilde{f} > 0 \\ \tilde{f}(x) \circledast \nabla(1 - \varepsilon(f) - 1.1\varepsilon) & \text{for } \tilde{f} < 0. \end{cases}$$

$\triangle$ means rounding upwards into the approximation screen S, ∇ rounding downwards. Possible overflow must be handled in a correct manner. For $\tilde{f}(x)=0$ the upper bound $\hat{f}(x)$ is given by the smallest positive floating-point number.

To obtain an upper bound in the coarser I/O-screen S^0, the following is done:

Cut the mantissa of $|\tilde{f}(x)|$ to the appropriate data format. If thereby a nonzero digit is lost and if $\tilde{f}(x)$ is positive, the upper bound in S^0 is the next larger floating-point number of this truncated value (overflow handling!).

Computation of a Point Approximation to $f(x)$

The computation is usually done in three steps:

- **argument reduction** into the reduction interval,
- **approximation** within the reduction interval,
- **result adaptation**.

Only Taylor polynomials are used as approximation functions. The polynomial coefficients are easy to compute. By adding some further terms the accuracy of the approximation may be arbitrarily increased.

2.1 Natural Logarithm $\log(X)$

Algorithm

Let $x = m \cdot B^e$ (>0, m mantissa, B basis, e exponent) be the input and $t := \dfrac{m \cdot c - 1}{m \cdot c + 1}$ the reduced argument. The appropriate constant c depends on the actual value of the mantissa m and is determined below. Then the logarithmic function is computed as

$$\log(x) = \log(m \cdot B^e) = \log(m) + e \cdot \log(B)$$

$$\approx t \cdot \sum_{n=0}^{N} \frac{2}{2n+1} t^{2n} - \log(c) + e \cdot \log(B).$$

The reduction interval $I_t = [-\gamma, \gamma]$ corresponds to the interval $I_x = \left[\dfrac{1-\gamma}{1+\gamma}, \dfrac{1+\gamma}{1-\gamma}\right]$ arround the zero of the logarithm. The numerator $m \cdot c - 1$ of t is for $x \in I_x$ exactly representable. Therefore, a bound for the *relative* error may be computed also for such arguments.

Argument Reduction to the Approximation Interval $I_t = [-\gamma, \gamma]$:

The interval $[B^{-1}, 1)$ is split into disjoint parts $[b_{i+1}, b_i)$, $\bigcup_i [b_{i+1}, b_i) = [B^{-1}, 1)$ each of which is mapped (using an appropriate constant c_i) into the approximation interval by the formula $\dfrac{m \cdot c_i - 1}{m \cdot c_i + 1}$. Thereby the breakpoints b_i and the corresponding reduction constants c_i are computed in the following way:

$$b_n = \left(\frac{1-\gamma}{1+\gamma}\right)^{2n+1}, \qquad c_n = \left(\frac{1+\gamma}{1-\gamma}\right)^{2n}.$$

The index n_0 of the smallest breakpoint b_{n_0} is determined by

$$n_0 := \min\left\{n \in \mathbb{N} \,\Big|\, n \geq \frac{\log(B)}{2 \cdot \log\dfrac{1+\gamma}{1-\gamma}} - \frac{1}{2}\right\} \quad (*).$$

b_{n_0} need not be stored because the interval $[b_{n_0}, b_{n_0-1})$ may be replaced by $[B^{-1}, b_{n_0-1})$. If the normalized mantissa m lies in the interval $[b_n, b_{n-1})$, $n \in \{1, 2, \ldots, n_0\}$, then the reduced argument $t := \dfrac{m \cdot c_n - 1}{m \cdot c_n + 1}$ lies within the approximation interval $[-\gamma, \gamma]$.

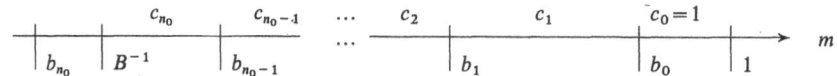

Figure 1. Reduction regions with corresponding reduction constants for $\log(x)$

Remarks:

- If the base B of the number system is 2^k, the normalization of the mantissa is done by shifts with respect to the base 2. In this case e means the exponent with respect to base 2. Only the interval $[0.5, 1)$ is managed by the argument reduction (n_0 is smaller, i.e. less constants have to be stored).
- The theoretical values of the breakpoints b_n and the reduction constants c_n are not representable on a computer. The actual values used must insure that the reduced arguments (afflicted with rounding errors) lie within the reduction inteval. The actual values of b_n and c_n may be computed using an interval arithmetic.

Error Bound for $\log(x)$:

$$\varepsilon(\log) \leq 1.1 \cdot \varepsilon(\text{app}) + 0.012 \cdot B \cdot \varepsilon + 3.6 \cdot \varepsilon + 6.9 \cdot n_0 \cdot \varepsilon$$
$$+ \frac{3.1}{\gamma}\{1 + \log(B)\} \cdot \varepsilon \qquad \text{for } \gamma < \frac{1}{8}.$$

This bound for the maximal relative error only depends on the screen accuracy ε, the base B, the approximation error $\varepsilon(\text{app})$ and the width of the reduction interval $[-\gamma, \gamma]$. The value of n_0 is given by formula $(*)$ for a fixed γ. The approximation error $\varepsilon(\text{app})$ is bounded by $\varepsilon(\text{app}) < \dfrac{1}{2N+3}\dfrac{\gamma^{2N+2}}{1-\gamma^2}$.

Example

To implement a logarithmic function with 100 hexadecimal digits ($\varepsilon(\log) < 16^{-99}$) the following may be done. Choosing $\gamma = 16^{-1.5}$ ($\Rightarrow n_0 = 11$ and $\log(B) = \log(2)$

< 0.7, for a binary normalized mantissa in the interval $[0.5, 1]$) the error estimation leads to

$$\varepsilon(\log) < 1.1 \cdot \varepsilon(\text{app}) + 3.8 \cdot \varepsilon + 76 \cdot \varepsilon + \frac{3.1}{\gamma} \{1 + 0.7\} \cdot \varepsilon$$

$$< 1.1 \cdot \varepsilon(\text{app}) + 418 \varepsilon < \varepsilon(100).$$

To satisfy this inequality an approximation degree of $N = 32$ is sufficient. In this case $\varepsilon(\text{app}) < 1.01 \dfrac{16^{-99}}{67} < 3.9 \cdot 16^{-101}$. Using a hexadecimal arithmetic with 103 mantissa digits for the approximation screen, i.e. $\varepsilon = 16^{-102}$, would give the final result

$$\varepsilon(\log) < 3.9 \cdot \varepsilon(102) + 1.7 \cdot \varepsilon(101) < \varepsilon(100).$$

Only the constants $\log(c_i)$ would be stored with a mantissa length of 103 hexadecimal digits.

Interval Evaluation of $\log(X)$:

The logarithm function is defined for $X = [\underline{x}, \bar{x}] > 0$ and is a monotonely increasing function. Therefore, $\log([\underline{x}, \bar{x}]) = [\log(\underline{x}), \log(\bar{x})]$.

2.2 Real Power Function X^Y

For the definition range and the interval version of this function of two variables see [4]. Only the error formula for dynamic accuracy is repeated from [5].

Algorithm

In almost all cases the formula $x^y = e^{y \cdot \log(x)}$ is used. See [5].

Error Bound for x^y:

$z = y \cdot \log(x)$ denotes the argument of the exponential function, and M is given by $M := \max\{|\log(\text{MINREAL})|, \log(\text{MAXREAL})\}$. M depends on the exponent range of the I/O-screen. A bound for the maximal relative error of the power function is

$$\varepsilon(\text{power}) < \{1.03 + \varepsilon(\log)\} \cdot M \cdot 1.02 + 1.01 \cdot \varepsilon(\exp),$$

where (without significant loss of generality) $M \cdot \varepsilon(z) < 0.01$ is assumed. The error from computing the logarithm is multiplied by the factor M. $\varepsilon(\exp)$ is given in [3].

Interval Evaluation of X^Y:

The computation of the interval bounds of the power function is described in [4].

2.3 Real Inverse Tangent Function arctan(X)

Algorithm

Defining $t := \dfrac{x - \tan(\alpha_i)}{1 + x \cdot \tan(\alpha_i)}$, where $x \cdot \tan(\alpha_i) > -1$, one finds for the input argument x:

$$\arctan(x) = \arctan\left(\frac{x - \tan(\alpha_i)}{1 + x \cdot \tan(\alpha_i)}\right) + \alpha_i = \arctan(t) + \alpha_i.$$

Within the approximation interval I_t, $\arctan(t)$ is approximated by

$$\arctan(t) \approx \sum_{k=0}^{N} \frac{(-1)^k}{2k+1} t^{2k+1}.$$

The constants $c_i \approx \tan(\alpha_i)$ for the argument reduction are chosen in such a manner that for the corresponding regions the reduced argument t lies in the reduction interval $I_t[-\gamma, \gamma]$. These constants are stored only with few mantissa digits whereas the constants $\alpha_i = \arctan(c_i)$ for the result adjustment are needed to the precision of the screen S.

Argument Reduction to the Approximation Interval $I_t = [-\gamma, \gamma]$:

For fixed γ, the breakpoints b_i are determined (and corresponding reduction constants c_i) for the interval splitting of $[0, 1/\gamma)$ into disjoint parts $[b_i, b_{i+1})$, $\bigcup_i [b_i, b_{i+1}) = [0, 1/\gamma)$, each of which is mapped via $\dfrac{x - \tan(\alpha_i)}{1 + x \cdot \tan(\alpha_i)} =: \dfrac{x - c_i}{1 + x \cdot c_i}$ into I_t. The interval $\left[\dfrac{1}{\gamma}, \text{MAXREAL}\right]$ is mapped into the reduction interval I_t using

$$\text{atan}(t) = \text{atan}\left(\frac{-1}{t}\right) + \frac{\pi}{2}.$$

Using $c_0 := 0$, $b_0 := 0$, and $b_1 := \gamma$ the reduction constant c_1 is given by $c_1 = \dfrac{b_1 + \gamma}{1 - \gamma b_1}$ (*) and the second breakpoint by $b_2 = \dfrac{\gamma + c_1}{1 - \gamma c_1}$ (**). In general the $(i-1)$st reduction constant is found by replacing all indizes in (*) by $i-1$. The endpoint of the i-th interval is found by replacing all indizes 1 in (**) by i. This process is stopped with the first breakpoint lying to the right of $\dfrac{1}{\gamma}$. For $|x|\left(<\dfrac{1}{\gamma}\right)$ in the interval $[b_i, b_{i+1})$, the reduced argument $t := \dfrac{|x| - c_i}{1 + |x| \cdot c_i}$ is a point of the approximation interval I_t.

Remarks:

● The breakpoints b_i and the reduction constants c_i are stored in a short data format. Only the constants $\alpha_i := \arctan(c_i)$ for the final adjustment of the result are stored in the full screen format.

● See also the second remark on the argument reduction of $\log(x)$.

Error bound for arctan (x):

The maximal relative error is bounded by

$$\varepsilon(\arctan) \leq 1.1 \cdot \varepsilon(\text{app}) + 0.014 \cdot B \cdot \varepsilon + 9 \cdot \varepsilon,$$

where the approximation error is bounded by $\varepsilon(\text{app}) < \dfrac{1.02}{2N+3} \gamma^{2N+2}$.

Interval Evaluation of arctan (X):

The inverse tangent function is monotonely decreasing in $\mathbb{R}$. The resulting interval is $\arctan([\underline{x}, \bar{x}]) = [\arctan(\underline{x}), \arctan(\bar{x})]$.

2.4 Real Inverse Cotangent Function arccot (X)

Algorithm

To compute the inverse cotangent the formula

$$\text{arccot}(y) = -\arctan(y) + \frac{\pi}{2}$$

is used. Because $\lim\limits_{x \to \infty} \arctan(x) = \dfrac{\pi}{2}$, cancellation must be avoided for large positive x. Therefore, in the interval $[T, \text{MAXREAL}]$ the approximation

$$\text{arccot}(x) = x \cdot \sum_{n=0}^{N} \frac{(-1)^n}{2n+1} x^{2n}, \quad x := \frac{1}{y}, \quad T > 8$$

is applied.

Error Bound for the Maximal Relative Error for arccot (x):

$$\varepsilon(\text{arccot}) < \max \left\{ \varepsilon(\text{app}) + 0.012 \cdot B \cdot \varepsilon + 3.25 \cdot \varepsilon, \frac{1}{\text{arccot}(T)} (4.8 \cdot \varepsilon + 1.6 \cdot \varepsilon(\arctan)) \right\},$$

where the approximation error is bounded by $\varepsilon(\text{app}) < \dfrac{1.02}{2N+3} \gamma^{2N+2}$. The breakpoint $T > 8$ is to be chosen in such a way that the two arguments of the maximum function are approximately equal.

Interval Evaluation of arccot (X):

The inverse cotangent is a monotonely decreasing function over $\mathbb{R}$. Therefore, $\text{arccot}([\underline{x}, \bar{x}]) = [\text{arccot}(\bar{x}), \text{arccot}(\underline{x})]$.

2.5 Real Inverse Hyperbolic Sine arsinh(X)

Algorithm

The following approximation is used in the interval $[-\gamma, \gamma]$:

$$\operatorname{arsinh}(x) \approx \sum_{n=0}^{N} a_n \cdot x^{2n+1}$$

with $a_0 = 1$ and $a_n = (-1)^n \dfrac{1 \cdot 3 \cdot 5 \cdot \ldots \cdot (2n-1)}{2 \cdot 4 \cdot 6 \cdot \ldots \cdot 2n} \cdot \dfrac{1}{2n+1}$ for $n \geq 1$.

For arguments not in $[-\gamma, \gamma]$, the formula

$$\operatorname{arsinh}(|x|) = \log(|x| + \sqrt{x^2 + 1}),$$

using the logarithm in combination with the squareroot function is applied.

Remark: If the intermediate results are not computable without overflow, the modified formula $\operatorname{arsinh}(|x|) \approx \log(2) + \log(|x|)$ may be used.

Error Bounds for arsinh(x):

With $|x| \leq \gamma < \frac{1}{8}$ the approximation error is bounded by

$$\varepsilon(\text{app}) < 1.02 |a_{N+1}| \cdot \gamma^{2N+2}.$$

As bound for the maximal relative error of the inverse hyperbolic sine one finds

$$\varepsilon(\text{arsinh}) < \max\left\{ \varepsilon(\text{app}) + 0.026 \cdot B \cdot \varepsilon + 2.2 \cdot \varepsilon, \left| \frac{\varepsilon(\log) + 1.1 \cdot \varepsilon(\text{sqrt}) + 2.2\varepsilon}{\operatorname{arsinh}(\gamma)} \right| \right\}.$$

$\varepsilon(\text{sqrt})$ is given in [3].

Interval Evaluation of arsinh(X):

Arsinh(x) is a monotonely increasing function in $\mathbb{R}$. Therefore, the resulting interval is $\operatorname{arsinh}([\underline{x}, \bar{x}]) = [\operatorname{arsinh}(\underline{x}), \operatorname{arsinh}(\bar{x})]$.

2.6 Real Inverse Hyperbolic Cosine arcosh(X)

For input arguments near 1, $\operatorname{arcosh}(x)$ is computed in the form $\operatorname{arcosh}(x) = f(x) \cdot \sqrt{x^2 - 1}$ using a function $f(x)$ which may be represented by a Taylor series with the center at 1. For other arguments the logarithmic representation is used.

Algorithm

Arcosh(x) is defined only for $x \geq 1$. In the range $1 \leq x < 1 + \gamma$ the representation

$$\operatorname{arcosh}(x) = P_N(t) \cdot \sqrt{x^2 - 1}$$

with

$$P_N(t) := \sum_{n=0}^{N} a_n \cdot t^n, \quad t := x - 1, \quad a_0 := 1$$

and

$$a_{n+1} := \frac{-n \cdot a_n}{2n+1}, \quad n \geq 0$$

is used. In the range $1 + \gamma \leq x < \textbf{MAXREAL}$ the formula

$$\text{acrosh}(x) = \log(x + \sqrt{x \cdot x - 1})$$

is applied.

Error Bounds for acrosh(x):

The approximation error is bounded by $\varepsilon(\text{app}) < \dfrac{1.02}{2N+3} \gamma^{2N+2}$.

Error bound for the maximal relative function error of arcosh(x):

$$\varepsilon(\text{arcosh}) < \max \left\{ \varepsilon(\text{app}) + 2.7 \cdot B \cdot \varepsilon + 1.2 \cdot \varepsilon(\text{sqrt}) + 2.2 \cdot \varepsilon, \right.$$

$$\left. \frac{1}{\text{arcosh}(1+\gamma)} (\varepsilon(\log) + 1.3 \cdot \varepsilon(\text{sqrt}) + 2.5 \cdot \varepsilon) \right\}$$

Interval Evaluation of arcosh(X):

arcosh(x) is in its definition range $x \geq 1$ a monotonely increasing function. For an input argument $X = [\underline{x}, \bar{x}]$ (≥ 1), the resulting interval is given by arcosh$(X) = [\text{arcosh}(\underline{x}), \text{arcosh}(\bar{x})]$.

2.7 Real Inverse Hyperbolic Tangent artanh(X)

This function is defined for arguments x with $|x| < 1$.

Algorithm

In the interval $[-\gamma, \gamma]$, $\gamma < 1/8$ the series approximation

$$\text{artanh}(x) \approx \sum_{n=0}^{N} \frac{1}{2n+1} x^{2n+1}, \quad |x| < \gamma$$

is used. Outside of this interval it is computed by

$$\text{artanh}(x) = 0.5 \cdot \log \frac{1+x}{1-x}.$$

Error Bound for the Maximal Relative Error of artanh(x):

The approximation error for arguments x lying in $[-\gamma, \gamma]$, $\gamma < 1/8$ is bounded by $\varepsilon(\text{app}) \leq \dfrac{1.02}{2N+3} \gamma^{2N+2}$.

Error bound for the maximal relative function error:

$$\varepsilon(\text{artanh}) < \max\left\{\varepsilon(\text{app}) + 0.012 \cdot B \cdot \varepsilon + 2.2 \cdot \varepsilon,\ 1.1 \cdot \varepsilon(\log) + 1.1 \cdot \varepsilon + \frac{3.8\,\varepsilon}{\text{artanh}(\gamma)}\right\}.$$

Interval Evaluation of artanh(X):

In its definition range $-1 < x < 1$ the inverse hyperbolic tangent is a monotonely increasing function. Therefore, $\text{artanh}([\underline{x}, \bar{x}]) = [\text{artanh}(\underline{x}), \text{artanh}(\bar{x})]$.

2.8 Real Inverse Hyperbolic Cotangent arcoth(X)

The definition range of $\text{arcoth}(x)$ is subdivided into two parts. Only arguments x with $|x| > 1$ valid.

Algorithm

For $|x| > \dfrac{1}{\gamma},\ \gamma < \dfrac{1}{8}$ the Taylor approximation

$$\text{arcoth}(x) \approx \sum_{n=0}^{N} \frac{1}{2n+1} \left(\frac{1}{x}\right)^{2n+1}$$

is used. Otherwise the formula

$$\text{arcoth}(x) = \frac{1}{2} \log \frac{x+1}{x-1}$$

is applied.

Upper Bound for the Maximal Relative Function Error of arcoth(x):

$$\varepsilon(\text{arcoth}) \leq \max\left\{3.8 \cdot \varepsilon \cdot \left(\log\left(\frac{\gamma+1}{\gamma-1}\right)\right)^{-1} + 1.01 \cdot (\varepsilon(\log) + \varepsilon),\right.$$

$$\left.\varepsilon(\text{app}) + 0.012 \cdot B \cdot \varepsilon + 2.2 \cdot \varepsilon\right\},$$

where in the regions $|x| > 1/\gamma,\ \gamma < 1/8$ the approximation error is bounded by

$$\varepsilon(\text{app}) \leq \frac{1}{2N+3} (1/\gamma)^{2N+2} \frac{1}{1-(1/\gamma)^2}.$$

Interval Evaluation of arcoth(X):

Within both subranges $x < -1$ and $x > 1$ the inverse hyperbolic cotangent function is monotonely decreasing. For any valid interval argument $X = [\underline{x}, \bar{x}]$ the resulting interval is given by $\text{arcoth}([\underline{x}, \bar{x}]) = [\text{arcoth}(\bar{x}), \text{arcoth}(\underline{x})]$.

For the following two routines it is assumed, that the input argument is afflicted with rounding errors. These routines are used to compute approxiations for the real and/or imaginary part of the complex valued functions described in § 3.

2.9 The Function $h(x)=\log(1+x)$

Algorithm to compute $h(x)=\log(1+x)$:

For $|x|<10^{-2}$, $h(x)$ is approximated by

$$h(x)\approx\sum_{n=0}^{N}\frac{2}{2n+1}\left(\frac{x}{2+x}\right)^{2n+1}.$$

Outside of this interval

$$h(x)\approx\log(1\oplus x)$$

is applied. In this case the argument $1\oplus x$ for the logarithm is explicitly computed.

Error Bound for the Maximal Relative Function Error of $h(x)$:

For arguments within the interval $-0.9<x,\ \tilde{x}<\text{MAXREAL}$ and for $\varepsilon(x)$
$<\dfrac{0.1}{\log(\text{MAXREAL}+1)}$ the following bound holds:

$$\left|\frac{h(x)-\tilde{h}(\tilde{x})}{h(x)}\right|\leq 4\cdot\varepsilon(x)+\varepsilon(h),$$

where $\varepsilon(h)$ is bounded by

$$\varepsilon(h)<\max\{1.1\,(\varepsilon(\text{app})+1.1\,\varepsilon+0.02\,B\varepsilon),\,102\varepsilon+1.1\,\varepsilon(\log)\}$$

and

$$\varepsilon(\text{app})<\frac{1.1}{2N+3}\cdot\left(\frac{1}{100}\right)^{2N+2}.$$

2.10 The Function $g(x)=\sqrt{1+x}-1$

Algorithm to compute $g(x)=\sqrt{1+x}-1,\ x\geq 0$:

For $x=0$ it is $g(x)=0$. For arguments in the range $0<x\leq 10^{-2}$ $g(x)$ is computed using the following Taylor series:

$$g(x)=x\sum_{k=0}^{\infty}(-1)^k\frac{1}{2}\left(\prod_{n=1}^{k}\frac{2n-1}{2n+2}\right)\cdot x^k=x\cdot\left\{\frac{1}{2}-\frac{1}{8}x+\frac{1}{16}x^2-+\ldots\right\}.$$

Within the interval $10^{-2}<x<\text{MAXREAL}$, the squareroot function is used:

$$\tilde{g}(x)=\sqrt[\tilde{}]{1\oplus x}\ominus 1.$$

Error Bound for the Maximal Relative Function Error of $g(x)$:

For $0\leq x\leq\text{MAXREAL}$ the function error is bounded by

$$\left|\frac{g(x)-\tilde{g}(\tilde{x})}{g(x)}\right|\leq 1.1\cdot\varepsilon(x)+\varepsilon(g),$$

where $\varepsilon(g)$ is bounded by

$$\varepsilon(g) < \max\left\{222 \cdot \varepsilon(\text{sqrt}) + 423\,\varepsilon,\ 1.3\left(\frac{1}{100}\right)^{N+1} + (2.5 + 0.005 \cdot B) \cdot \varepsilon\right\}.$$

$\varepsilon(x)$ denotes the maximal relative error of the input argument $\tilde{x}$.

3. Complex Interval Functions

3.0 Computation of Bounds for a Complex Standard Function

A complex interval is given by its lower left and its upper right corner, i.e. rectangle arithmetic is assumed.

In contrast to the real interval standard functions, the minimum and maximum of the real and imaginary part are not usually found at the corners of the input interval or at any other floating-point number on the boundary of the input interval. The extreme values are, in general, found as intersection points of extremal curves with the boundary of the input interval. The location of extrema differs for the real and imaginary part. To get an approximation of an extreme value, first the location of this extremum has to be found. The coordinates of such a point are given by formulas and are not exactly representable on the computer. However, the extrema are found on the boundary of the input interval, and therefore at least one of the two coordinates is exactly known. The formulas for the point approximations use this coordinate.

General Program Description for Computing a Complex Interval Standard Function $f(Z) = f(X + iY)$

Main routine:

1) Error stop if an illegal argument lies in the input interval.
2) If points of the branch cut belong to the input interval, the flag CUT is set. If the input argument is a point argument, the flag POINT_ARGUMENT is set.
3) Call the subroutine for the computation of approximations for the bounds of the resulting real part interval (input parameters are X, Y and the flags CUT and POINT_ARGUMENT; output parameters are approximations RLB of the lower bound and RUB of the upper bound of the real part).
4) Directed rounding of the approximation screen values RLB and RUB into the I/O-screen.
5) Subroutine call for the computation of approximations for the bounds of the resulting imaginary part interval (input parameters are X, Y, CUT and POINT_ARGUMENT; output parameters are approximations of the lower bound ILB and of the upper bound IUB of the imaginary part).
6) Directed rounding of the approximation screen values ILB and IUB into the I/O-screen.
7) Combination of the bounds for the real and imaginary parts to get the lower left and the upper right corner of the resulting complex interval.

Subroutine for Computing the Real Part:

POINT_ARGUMENT is set: computation of a point approximation of the function value and assignment to RLB and RUB; return.

POINT_ARGUMENT is not set: If CUT is set special interval logic for intervals affected by the cut is performed, otherwise normal interval logic is executed. As result the locations of the extrema of the real part function are given. Using these locations, approximations to RLB and RUB are computed (point approximation routine is called twice); return.

The subroutine for the computation of the imaginary part is analogous.

Remark: If continuations over the cuts are used, inclusion monotonicity no longer holds, i.e.

$$Z_1 \subset Z_2 \not\Rightarrow f(Z_1) \subset f(Z_2).$$

Inclusion monotonicity holds for all intervals lying in the range corresponding to the principal value of the complex function.

3.1 Absolute Value of a Complex Interval abs(Z)

Definition: $\mathrm{abs}(z) = \sqrt{x^2 + y^2}$.

Algorithm

If the exponents of x and y differ by more than the desired number of mantissa digits, then the approximation

$$\mathrm{abs}(z) = \mathrm{abs}(x, y) \approx \max\{|x|, |y|\}$$

may be applied. Otherwise

$$\mathrm{abs}(z) = B^{\frac{ex+ey}{2}} \cdot \sqrt{(mx)^2 \cdot B^{ex-ey} + (my)^2 \cdot B^{-ex+ey}}, \quad ex - ey \quad \text{even}$$

or

$$\mathrm{abs}(z) = B^{\frac{ex+ey-1}{2}} \cdot \sqrt{B \cdot \{(mx)^2 \cdot B^{ex-ey} + (my)^2 \cdot B^{-ex+ey}\}}, \quad ex - ey \quad \text{odd}$$

is used. mx, my are the mantissa of x, y and ex, ey their exponents. Exponent manipulations are shifts with respect to the base B.

Interval Evaluation of abs(Z):

With $\underline{z} := \min|x| + i \cdot \min|y|$ and $\bar{z} := \max|x| + i \cdot \max|y|$ the resulting interval is given by $\mathrm{abs}(Z) = [\mathrm{abs}(\underline{z}), \mathrm{abs}(\bar{z})]$.

Error Bound for abs(z):

The maximal relative error is bounded by

$$\varepsilon(\mathrm{abs}) < \max\{B \cdot \varepsilon, 1.1 \cdot \varepsilon + \varepsilon(\mathrm{sqrt})\}.$$

3.2 Argument of a Complex Interval arg(Z)

Algorithm

With $g(x, y) := \arctan\left(\dfrac{y}{x}\right)$ one finds

$$\overline{\arg}(x, y) = \begin{cases} g(x, y) & z \text{ in quadrant } I \text{ or } IV \\[4pt] \dfrac{\pi}{2} & z \text{ on positive imaginary axis} \\[4pt] g(x, y) + \pi & z \text{ in } II \\[4pt] g(x, y) - \pi & z \text{ in } III \\[4pt] \dfrac{3}{2}\pi & z \text{ on negative imaginary axis} \end{cases}$$

Here $\overline{\arg}$ means the canonical principal value of the argument function with $-\pi < \overline{\arg}(x, y) \leq \pi$. The function $g(t) = g(x, y) = \arctan\left(\dfrac{y}{x}\right)$, $t = \dfrac{y}{x}$, is computed by $g(|t|) = \arctan\left(\left|\dfrac{y}{x}\right|\right)$ for $|t|$ small and by $g(|t|) = \arctan\left(-\left|\dfrac{x}{y}\right|\right) + \dfrac{\pi}{2}$ for $|t|$ large.

Interval Evaluation of arg(Z):

The point interval $Z = [(0, 0), (0, 0)]$ leads to an error. For intervals which lie across the negative real axis, the function values of that part of the interval which lies below the x-axis are computed on a second Riemannian surface. The minimum of the resulting interval is found in the upper right corner of Z and the maximum ($> \pi$) is found in the lower right corner of the input interval. For intervals which do not cross the negative real axis, figure 2 gives the positions where the extreme values are found.

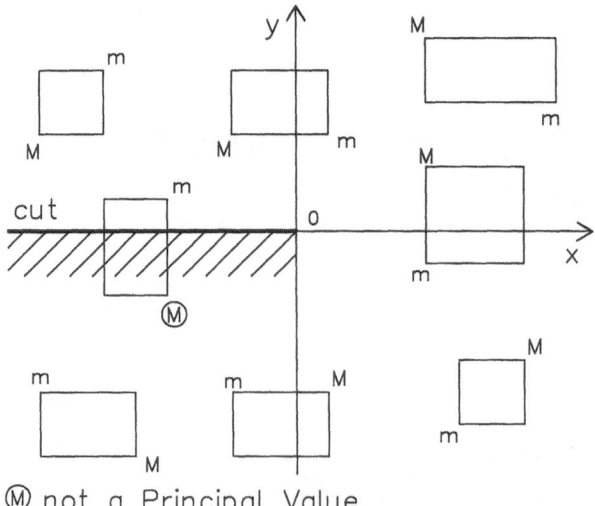

Figure 2. Position of extreme values of arg(Z)

Error Bound for $\arg(z)$:

The maximal relative error when computing $\arg(z)$ is bounded by

$$\varepsilon(\mathbf{arg}) < 6.4 \cdot \varepsilon + 1.3 \cdot \varepsilon(\mathrm{arctan}).$$

3.3 Complex Natural Logarithm $\log(Z)$

Definition: $\mathrm{Re}(\log(x, y)) = \log\sqrt{x^2 + y^2}$, $\mathrm{Im}(\log(x, y)) = \arg(x, y)$.

Algorithm

Within the region $|1 - x^2 - y^2| < \gamma\left(\approx \dfrac{1}{100}\right)$

$$\mathrm{Re}(\overline{\log}(x, y)) = \tfrac{1}{2}\log(x^2 + y^2) = \tfrac{1}{2}h(x^2 + y^2 - 1)$$

with $h(x) = \log(1 + x)$ is used. Outside this region,

$$\mathrm{Re}(\overline{\log}(x, y)) = \tfrac{1}{2}\log(x^2 + y^2) \quad \text{is used.}$$

The imaginary part is computed by

$$\mathrm{Im}(\overline{\log}(x, y)) = \overline{\arg}(x, y), \quad -\pi < \overline{\arg}(z) \leq \pi.$$

Remark: Modification of the formula for the real part:

$$\mathrm{Re}(\overline{\log}(x, y)) = \tfrac{1}{2}\log(x^2 + y^2)$$
$$= \tfrac{1}{2}\{(ex + ey) \cdot \log(B) + \log(mx^2 \cdot B^{ex-ey} + my^2 \cdot B^{-(ex-ey)})\},$$

where mx, my are the mantissas and ex, ey the exponents of x and y.
If the difference of the exponents of x and y is large enough $\mathrm{Re}(\overline{\log}(x, y)) = \log(\max\{|x|, |y|\})$ may be used.

Branch Cut for $\log(z)$:

As branch cut the negative real axis is chosen. Intervals lying across the cut are continued from the upper into the lower half plane. The function values for the part of the input interval lying in the lower half plane are computed on a second Riemannian surface using $\log(z) = \overline{\log}(z) + \pi$. Illustrating figures are given in [4].

Interval Evaluation of $\log(Z)$:

If $(0, 0) \in Z$, error stop.

Interval evaluation of the real part Re(log(Z)):

With $Z := [xlb, xub] + i[ylb, yub]$,

$$\underline{x} := \begin{cases} xlb, & \text{for } 0 < xlb \le xub \\ |xub|, & \text{for } xlb \le xub < 0 \\ 0, & \text{for } xlb \le 0 \le xub, \end{cases}$$

$$\bar{x} := \begin{cases} xub, & \text{for } 0 < xlb \le xub \\ |xlb|, & \text{for } xlb \le xub < 0 \\ \max\{|xlb|, xub\}, & \text{for } xlb \le 0 \le xub \end{cases}$$

one finds $\mathrm{Re}(\overline{\log}(Z)) = \frac{1}{2}\log[\underline{x}^2 + \underline{y}^2, \bar{x}^2 + \bar{y}^2]$. $\underline{y}$ and $\bar{y}$ are determined similarly.

Interval evaluation of the imaginary part Im(log(Z)):

The interval evaluation of $\mathrm{Im}(\log(Z))$ is done in the same way as the interval evaluation of the argument function.

Error Bounds

Error bound for the maximal relative error of Re(log(z)):

$$\varepsilon(\mathrm{Re}(\log)) = \max\left\{ 4.1\,\varepsilon + 0.6\,\varepsilon(\log) + \frac{1.32}{\log(1+\gamma)}\,\varepsilon, \quad 2.2\,\varepsilon + \varepsilon(h) \right\}.$$

$\varepsilon(h)$ is the bound for the implemented function $h(x) = \log(1+x)$.

Error bound for the maximal relative error of Im(log(z)):

$$\varepsilon(\mathrm{Im}(\log)) = \varepsilon(\arg).$$

3.4 Complex Arcsine Function arcsin(Z)

Logarithmic representation:

$$\arcsin(z) = w = -i \cdot \log(iz \pm \sqrt{1 - z^2})$$

i.e.

$$\mathrm{Re}(\arcsin(z)) = \arg(iz \pm \sqrt{1 - z^2}) + 2k\pi \quad \text{and}$$

$$\mathrm{Im}(\arcsin(z)) = -\log(|iz \pm \sqrt{1 - z^2}|),$$

where the sign in both formulas has to be the same.

Branch Cut for arcsin(z):

The complex plane is cut along the real axis from $-\infty$ to -1 and from 1 to ∞. The end points -1 and 1 are winding points. The mapping given by arcsin is not a conformal mapping at these points. When approaching the cut $(-\infty, -1)$ from above, the imaginary part is greater than zero, when approaching it from below, it is negative. The real part tends to the value $-\dfrac{\pi}{2}$ from both sides. When approaching the second cut $(1, \infty)$ from above, the imaginary part is positive, when approaching it from below, it is negative. The real part tends to $\dfrac{\pi}{2}$ in both cases.

Continuation (counterclockwise over the cuts to assure $\arcsin(-Z) = -\arcsin(Z)$ also for such intervals) to another Riemannian surface is only done for intervals Z lying across a cut with $1, -1 \notin Z$. For intervals containing one or both winding points only principal values are included.

Formulas for computation of the real part $\mathrm{Re}(\arcsin(z))$:

$$\mathrm{Re}(\overline{\arcsin(z)}) = \arcsin\left(\frac{2 \cdot x}{\sqrt{(x+1)^2 + y^2} + \sqrt{(x-1)^2 + y^2}}\right)$$

is used in the range $x^2 - y^2 < 0.5$ and

$$\mathrm{Re}(\overline{\arcsin(z)}) = \frac{\pi}{2} - \arcsin\sqrt{\frac{y^2 - x^2 + 1 + \sqrt{(x^2 + y^2 - 1)^2 + 4y^2}}{y^2 + x^2 + 1 + \sqrt{(x^2 + y^2 - 1)^2 + 4y^2}}}$$

in the range $x^2 - y^2 \geq 0.5$. In the last case the numerator of the radicand is evaluated as

$$\begin{cases} 2y^2 + |x^2 + y^2 - 1| \cdot g(s^2) & \text{for } x^2 + y^2 \geq 1 \\ 2 - 2x^2 + |x^2 + y^2 - 1| \cdot g(s^2) & \text{for } x^2 + y^2 < 1 \end{cases}$$

with $g(s) := \sqrt{1 + s} - 1$ and $s := \dfrac{2y}{x^2 + y^2 - 1}$.

If the function value for a point on another Riemannian surface is to be computed, π is added or subtracted from $\mathrm{Re}(\overline{\arcsin(z)})$.

Formula for the computation of the imaginary part $\mathrm{Im}(\arcsin(z))$:

The formula

$$\mathrm{Im}(\overline{\arcsin(z)}) = \log(t + \sqrt{t^2 - 1})$$

with

$$t(z) = t(x, y) = 0.5\sqrt{(|x| + 1)^2 + y^2} + 0.5\sqrt{(|x| - 1)^2 + y^2}$$

is used in the range $t > 1 + \gamma$ ($\gamma \approx 0.01$). Outside this range, i.e. for $1 \leq t \leq 1 + \gamma$ the following modifications are made:

$$\mathrm{Im}(\overline{\arcsin(z)}) = h(s - 1) \quad \text{with} \quad h(x) = \log(1 + x)$$

and

$$s - 1 = r + \sqrt{2r} \cdot \sqrt{1 + r/2},$$

$$r = t - 1 = |x| - 1 + (|x| + 1) \cdot g\left(\left(\frac{y}{|x| + 1}\right)^2\right) + \sqrt{(|x| - 1)^2 + y^2}$$

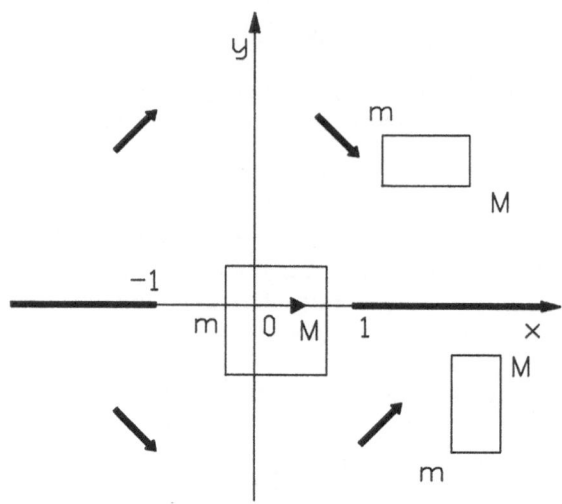

Figure 3. Monotonicity and position of the extreme values of Re(arcsin(Z))

and

$$g(x) = \sqrt{1+x} - 1.$$

If the function value for a point z on another Riemannian surface is to be computed, the sign of $\mathrm{Im}\,(\overline{\arcsin(z)})$ has to be inverted.

Interval Evaluation of arcsin(Z)

Interval evaluation of the real part Re$(\overline{\arcsin}(Z))$:

Using the partial derivatives of Re$(\overline{\arcsin}(x, y))$ with respect to x and y one finds the monotonicity behaviour as shown in figure 3. The function increases in the directions given by the arrows. The letter m indicates the position of a minimum, M the position of a maximum of the function.

In the right half plane the function is positive whereas in the left half plane it is negative.

Interval evaluation of the imaginary part Im$(\overline{\arcsin}(Z))$:

Figure 4 shows the monotonicity behaviour of the imaginary part of the Principal Value $\overline{\arcsin}(z)$.

Considering the isolines (parts of ellipses) of Im$(\overline{\arcsin}(z))$, it is possible to decide which of the points m?, M? have to be chosen.

Error Bounds

$\varepsilon(g)$ denotes an error bound for the maximal relative error of $g(x) := \sqrt{1+x} - 1$ and $\varepsilon(h)$ of $h(x) := \log(1+x)$.

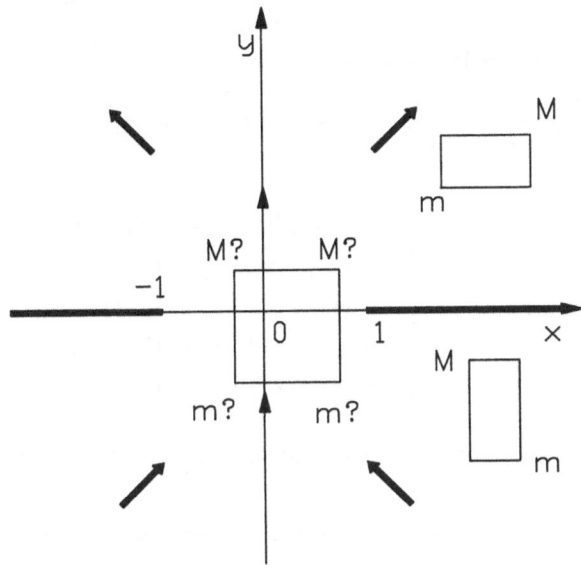

Figure 4. Monotonicity and position of the extreme values of $\mathrm{Im}(\overline{\mathrm{arcsin}}(Z))$

Error bound for the maximal relative error of $\mathrm{Re}(\mathrm{arcsin}(z))$:

$$\varepsilon(\mathrm{Re}(\mathrm{arcsin})) < \max\{6.3\,\varepsilon + 2\varepsilon(\mathrm{sqrt}) + \varepsilon(\mathrm{arcsin}),$$
$$9.6\,\varepsilon + 0.4\varepsilon(g) + 0.4\varepsilon(\mathrm{sqrt}) + 1.1\,\varepsilon(\mathrm{arcsin})\}.$$

For $\varepsilon(\mathrm{arcsin})$ see [2].

Error bound for the maximal relative error of $\mathrm{Im}(\mathrm{arcsin}(z))$:

$$\varepsilon(\mathrm{Im}(\mathrm{arcsin})) < \max\left\{\frac{2.2}{\mathrm{arcosh}(1+\gamma)} + \varepsilon(\log), 8.6\varepsilon + 1.6\cdot\max\{4.5\varepsilon + \varepsilon(g), \varepsilon(\mathrm{sqrt})\}\right.$$
$$\left. + 1.3\cdot\varepsilon(\mathrm{sqrt}) + \varepsilon(h)\right\}.$$

3.5 Complex Inverse Tangent $\mathrm{arctan}(Z)$

Definition of real and imaginary part of the inverse tangent:

$$\mathrm{Re}(\mathrm{arctan}(z)) = \frac{-1}{2}\,\mathrm{arg}\left(\frac{i+z}{i-z}\right), \quad \mathrm{Im}(\mathrm{arctan}(z)) = \frac{-1}{2}\,\log\left|\frac{i+z}{i-z}\right|.$$

Branch Cut of the Inverse Tangent:

The cuts are on the imaginary axis from $-\infty\cdot i$ to $-i$ and from i to $i\cdot\infty$. The points $\pm i$ are singularities and the function is not defined there.

Algorithm

Formula for the real part Re(arctan(z)):

With $g(x, y) := \frac{1}{2} \arctan \frac{2 \cdot x}{1 - x^2 - y^2}$, the real part function $\mathrm{Re}(\overline{\arctan}(z))$, $z = x + i \cdot y$ is given by

$$\mathrm{Re}(\overline{\arctan}(z)) = \begin{cases} g(x, y) + \pi/2, & z \in I \cup IV, \quad |z| > 1 \\ g(x, y), & |z| < 1 \\ g(x, y) - \pi/2, & z \in II \cup III, \quad |z| > 1 \\ -\pi/4, & |z| = 1, \quad x < 0 \\ \pi/4, & |z| = 1, \quad x > 0. \end{cases}$$

z lies in the quadrants indicated by roman numbers. For z not on the imaginary axis one obtains

$$\mathrm{Re}(\overline{\arctan}(x, y)) = \mathrm{Re}(\overline{\arctan}(x, -y)) = \mathrm{Re}(\overline{\arctan}(x, |y|)).$$

For z on the imaginary axis

$$\mathrm{Re}(\overline{\arctan}(0, y)) = 0, \quad -1 < y < 1 \quad \text{and}$$
$$\mathrm{Re}(\overline{\arctan}(0, y)) = -\mathrm{Re}(\overline{\arctan}(0, -y))$$

is used. If the function value has to be determined for an argument on the extremal curve $y^2 - x^2 = 1$, it is computed in the form

$$\mathrm{Re}(\overline{\arctan}(\sqrt{y^2 - 1}, y)) = \frac{1}{2} \arctan \left(\sqrt{\frac{1}{y^2 - 1}} \right).$$

For $y = 1$ the formula

$$\mathrm{Re}(\overline{\arctan}(z)) = \frac{1}{2} \arctan \left(\frac{-2}{x} \right)$$

is used. If z lies in quadrant II and the function value has to be computed on another Riemannian surface, π is added to $\mathrm{Re}(\overline{\arctan}(z))$. If z lies in quadrant IV, π is subtracted from $\mathrm{Re}(\overline{\arctan}(z))$. In the first case the function values are $> \pi/2$, in the second case $< -\pi/2$. The symmetry relation

$$\mathrm{Re}(\arctan(X, Y)) = -\mathrm{Re}(\arctan(-X, Y))$$

holds in every case.

Formula for the imaginary part Im(arctan(z)):

The imaginary part $\mathrm{Im}(\overline{\arctan}(z))$ is computed by

$$\mathrm{Im}(\overline{\arctan}(z)) = \frac{1}{4} \log \left(\frac{x^2 + (1 + y)^2}{x^2 + (1 - y)^2} \right), \quad \text{for } z \neq \pm i.$$

If the argument of the real logarithm is close to 1, the formula is modified.

In this case

$$\mathrm{Im}\,\overline{(\arctan(z))} = \frac{1}{4}\,h\!\left(\frac{4y}{x^2+(1-y)^2}\right).$$

where $h(x):=\log(1+x)$ is used. If the argument for the imaginary part function lies on an extremal curve one gets

$$\mathrm{Im}\,\overline{(\arctan(z))} = \mathrm{Im}\,\overline{(\arctan(x,\sqrt{x^2+1}))} = \frac{1}{4}\,\log\!\left(\frac{1}{2\cdot(\sqrt{1+x^2}-1)}\right)$$

$$= \frac{1}{4}\,\log\!\left(\frac{1}{2\cdot g(x^2)}\right), \quad \text{with } g(t):=\sqrt{1+t}-1.$$

Interval Evaluation of arctan(Z)

Interval evaluation for the real part $\mathrm{Re}(\arctan(z))$:

Figure 5 shows how the interval logic for the real part $\mathrm{Re}(\arctan(Z))$ can be implemented. The following abbreviations are used:

$Z=(X,\,Y)$ complex input interval, $X=[\underline{x},\,\bar{x}]$ real part interval, $Y=[\underline{y},\,\bar{y}]$ imaginary part interval of Z, $R=[\underline{r},\,\bar{r}]$ real part of resulting complex interval, E: $y=+\sqrt{1+x^2}$ extremal curve, $e=\sqrt{y^2-1}$, $u(x,\,y)=\mathrm{Re}\,\overline{(\arctan(x,\,y))}$, $F \hat{=}$ flag CUT, $\overline{II} \hat{=}$ quadrant II including negative real and positive imaginary axis.

To find the extreme values for intervals which do not belong to the upper half plane, symmetry relations are used:

In quadrant II

$$\mathrm{Re}(\arctan(X,\,Y)) = \mathrm{Re}(\arctan(-X,\,Y)),$$

in quadrants III or IV

$$\mathrm{Re}(\arctan(Z)) = -\mathrm{Re}(\arctan(-Z)) \quad (-Z \text{ lies in } I \text{ or } II)$$

and for intervals which intersect the real axis

$$\mathrm{Re}(\arctan(Z)) = \mathrm{Re}(\arctan(X,\,|Y|)).$$

Interval evaluation of the imaginary part $\mathrm{Im}\,\overline{(\arctan(Z))}$:

Similar diagrams as given for $\mathrm{Re}(\arctan(Z))$ may be developed. Case selection may be gathered from the following. For input intervals lying in quadrant I, figure 6 shows all possible constellations. The position of the intervals with respect to the extremal curves is important.

Intervals in other quadrants are handled via

$$\mathrm{Im}(\arctan(X,\,Y)) = -\mathrm{Im}(\arctan(X,\,-Y)),$$
$$\mathrm{Im}(\arctan(-X,\,Y)) = \mathrm{Im}(\arctan(X,\,Y))$$

or

$$\mathrm{Im}(\arctan(-X,\,-Y)) = -\mathrm{Im}(\arctan(X,\,Y)).$$

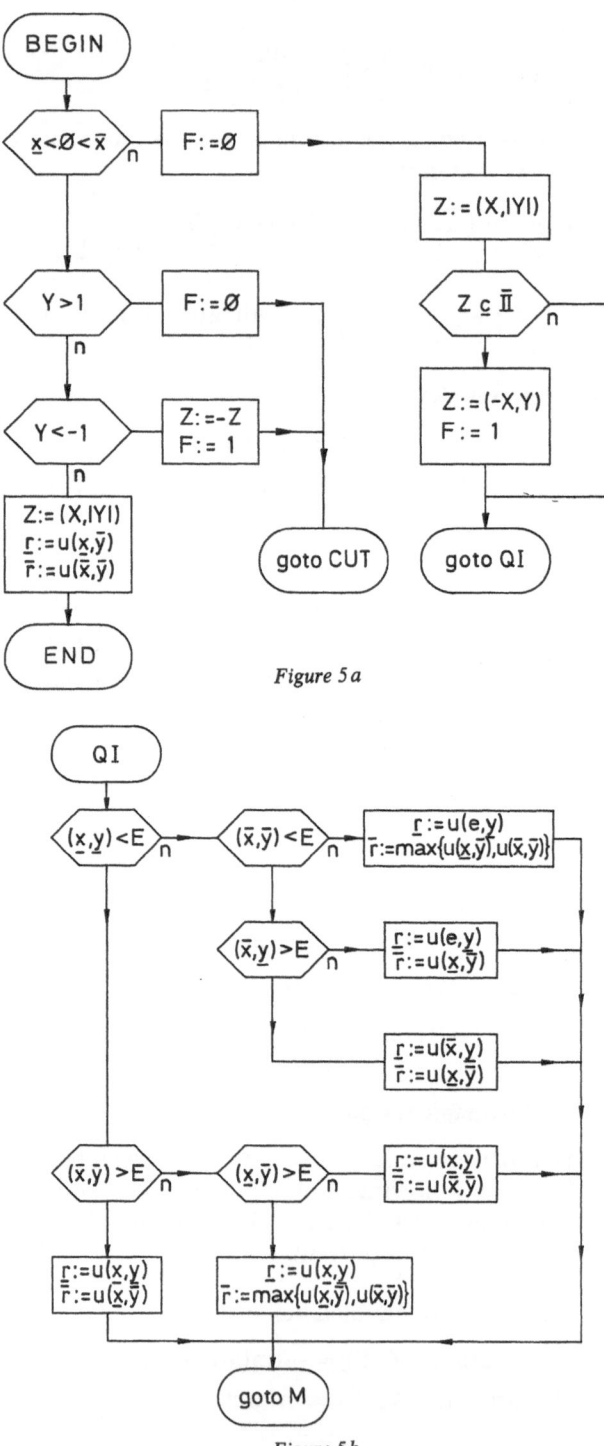

Figure 5a

Figure 5b

Figure 5. Interval logic of Re(arctan(Z))

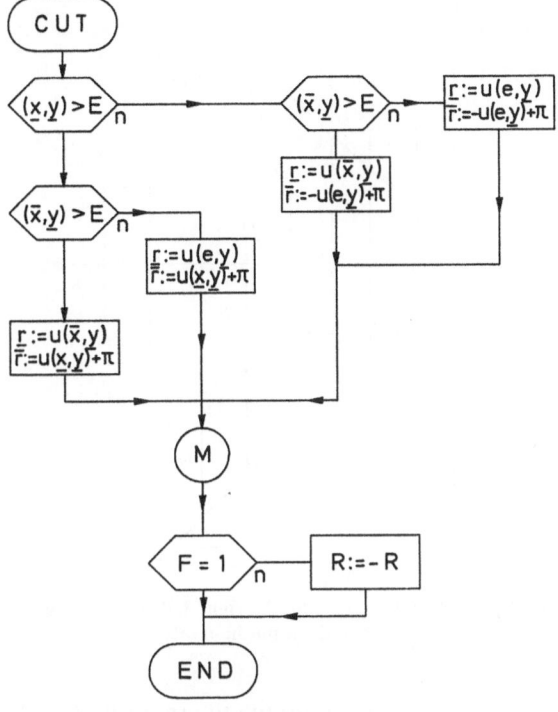

Figure 5c

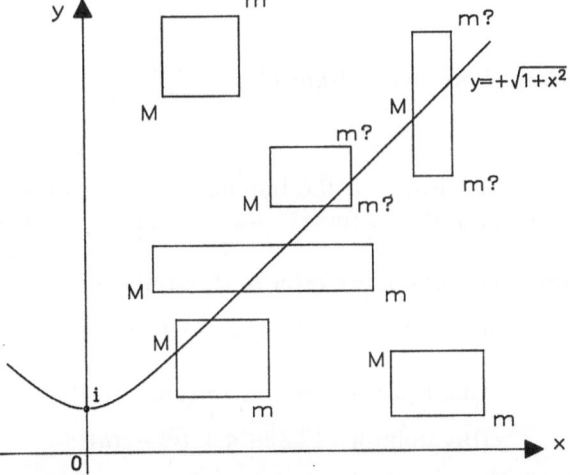

Figure 6. Interval evaluation of $\text{Im}(\overline{\arctan(Z)})$ for intervals in quadrant I

W. Krämer:

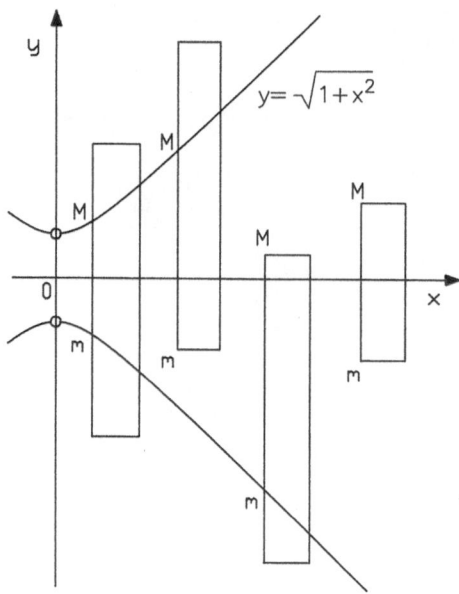

Figure 7. Interval evaluation of $\mathrm{Im}(\overline{\arctan}(Z))$ in the right half plane with 0 inside the imaginary part of the input interval

In particular, intervals Z which have points in common with the left half plane are replaced by $Z=(|X|, Y)$. Intervals in quadrant IV are reflected at $(0, 0)$ and the sign of the result is changed. Finally, for intervals in the right half plane with 0 in their imaginary part, the lower bound of the result is found in quadrant IV and the upper bound in quadrant I. Four subcases are to be kept in mind (see figure 7). The left boundary may intersect the extremal curve

- twice,
- within quadrant I,
- only in quadrant IV,
- neither in quadrant I nor in quadrant IV.

Error Bounds

To check whether z lies inside, on the boundary or outside of the unit circle, the exact scalar product of the vectors $(1, -x, -y)$, $(1, x, y)$ is used.

Error bound for the maximal relative error of $\mathrm{Re}(\arctan(z))$:

$$\varepsilon(\mathrm{Re}(\arctan)) < 8.6\varepsilon + 1.7\varepsilon(\mathrm{sqrt}) + 1.2\varepsilon(\arctan).$$

Error bound for the maximal relative error of $\mathrm{Im}(\arctan(z))$:

$$\varepsilon(\mathrm{Im}(\arctan)) < 12.8\varepsilon + 4.4\varepsilon(g) + \varepsilon(h).$$

It is assumed that $\frac{1}{4}$ is exactly representable. $\varepsilon(g)$ and $\varepsilon(h)$ are given in § 3.9 and § 3.10.

References

[1] IBM High-Accuracy Arithmetic Subroutine Library (ACRITH), Program Description and User's Guide. SC 33-6164-02, 3rd Edition, April 1984.

[2] Braune, K.: Hochgenaue Standardfunktionen für reelle und komplexe Punkte und Intervalle in beliebigen Gleitpunktrastern. Dissertation, Universität Karlsruhe, 1987.

[3] Braune, K.: Standard Functions for Real and Complex Point and Interval Arguments with Dynamic Accuracy. This Volume.

[4] Braune, K., Krämer, W.: High-Accuracy Standard Functions for Real and Complex Intervals. Computerarithmetic: Scientific Computation and Programming Languages. E. Kaucher, U. Kulisch, Ch. Ullrich (ed.), Teubner, Stuttgart, 81 – 114, 1987.

[5] Krämer, W.: Inverse Standardfunktionen für reelle und komplexe Intervallargumente mit a priori Fehlerabschätzungen für beliebige Datenformate. Dissertation Universität Karlsruhe, 1987.

[6] Kulisch, U., Miranker, W. L.: Computer Arithmetic in Theory and Practice. Academic Press, New York, 1981.

Dr. Walter Krämer
Institut für Angewandte Mathematik
Universität Karlsruhe
Kaiserstrasse 12
D-7500 Karlsruhe
Federal Republic of Germany

Computing, Suppl. 6, 213–224 (1988)

Computing
© by Springer-Verlag 1988

Inclusion Algorithms with Functions as Data

H. J. Stetter, Vienna

Abstract — Zusammenfassung

Inclusion Algorithms with Functions as Data. An algorithm can only use finite information about data, hence a large set of data functions may look alike to a given inclusion algorithm which would then have to include the corresponding result set. It is examined how this dilemma may be overcome by the specification of function data through arithmetic expression strings.

Einschließungsalgorithmen mit Funktionen als Daten. Ein Algorithmus kann nur endliche Dateninformation benützen; deshalb kann eine große Menge von Datenfunktionen für einen vorliegenden Einschließungsalgorithmus gleich aussehen und dieser müßte die entsprechende Ergebnismenge einschließen. Es wird untersucht, wie dieses Dilemma überwunden werden kann, indem man Funktionsdaten durch Arithmetic Expression Strings spezifiziert.

1. Numerical Analysis vs. Numerical Algebra

When we look at existing numerical software with automatic result verification, we find a relative wide variety available for problems for Numerical Algebra but only a few packages for problems of Numerical Analysis. In this paper, we wish to exhibit some fundamental reasons for this discrepancy and to suggest directions for research and for the design of inclusion algorithms in Numerical Analysis.

Let us, at first, clarify one distinction between Numerical Algebra and Analysis:

In *Numerical Algebra*, we consider algorithms whose data consist of *finite sets of real numbers*. In contrast, the algorithms in *Numerical Analysis* have also *functions* (from reals to the reals) as data. Note that it is the occurence of functions as input data of algorithms which marks the distinction, not the occurence of functions as such: The evaluation of a trigonometric polynomial (or equivalently the FFT-algorithm) is part of Numerical Algebra in our sense since the trigonometric functions are *constants* of the algorithm.

Examples:

Typical *Numerical Algebra* problems are linear systems, algebraic eigenvalue problems, linear programs, but also the evaluation of arithmetic expressions which involve fixed functions as operators.

Typical *Numerical Analysis* problems are

- Zero of a function $f(y) = 0$
- Integration $y = \int_a^b f(t)\,dt$
- Differentiation $y(x) = \dfrac{d}{dx} f(x)$
- Differential Equations $y''(t) + f_1(t)\,y'(t) + f_2(t)\,y(t) = f_3(t)$

In each case, f or f_i signify the data functions of the problem while y signifies the result (number or function). □

This definition of the borderline makes sense: In Numerical Algebra the information in the data may be fully entered into the algorithm while the information contained in a general real function is infinite and can enter a numerical algorithm (in the real domain) only via finitely many functionals. Thus the information which is used in some algorithm of Numerical Analysis may, generally, come from some function out of a infinite set of data functions which are equivalent and indistinguishable for the particular algorithm. It is this lack of a bi-unique relation between the information used by an algorithm and by the mathematically specified problem which gives a different perspective to the design of algorithms with result verification in Numerical Analysis.

In the following, we shall at first study the data-result relationship in Numerical Analysis in more detail and then point out how a firm basis for inclusion algorithms may be regained in Numerical Analysis. Consequences of these considerations will be exhibited.

2. The Data-Result Relation in Numerical Analysis

The typical structure of a computational algorithm in Numerical Analysis is the following [1] (cf. a quadrature or an o.d.e. algorithm):

The data function(s) are assumed to be given in the form of procedures which permit evaluation at specified argument values. The main algorithm succesively retrieves values of the data function (the integrand, the right hand side of the differential equation) and performs "algebraic" operations with them and with other numerical data (coefficients, stepsize) to generate one or several result values (the value of the integral, the values of the solution on a specified grid). Result functions may be generated through interpretation of result values as parameters in a fixed "Ansatz".

Assume that N values of the (only) data function f have been retrieved and L result values have been generated which represent an approximate result function $\tilde{y}$, for a mathematical problem which associates a unique exact result function y in some function space R with the data function f. Then we have

---

[1] This presentation has been strongly patterned after the ideas of Traub and Wosniakowski (e.g. [8])

the following relational diagram for some algorithm M which "solves" this problem:

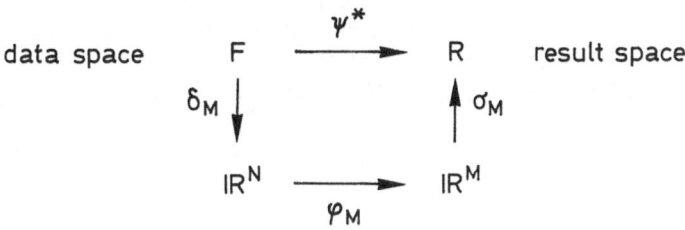

δ_M denotes the data function evaluation,
φ_M the computation proper,
σ_M the functional interpretation,
ψ^* the mathematical solution mapping.

In practically all natural situations, the mathematical solution mapping of the problem will have its domain in an infinite-dimensional function space F. Therefore, the image

$$\delta = \delta_M(f) \in \mathbb{R}^N$$

of our particular data function f will also be the image of an infinite set $D_M(f)$ of other, *equivalent*, data functions in F which *look like f* to our algorithm M and for which M will produce one and the same approximate result $\tilde{y} \in R$.

However, the exact results $\psi^*(g)$, $g \in D_M(t)$, will generally not at all coincide with $y^* = \psi^*(f)$ but form an infinite set of "equivalent results" $R_M(f)$. Since our algorithm M cannot tell f from its brothers in $D_M(f)$, it certainly cannot distinguish between results in $R_M(f)$. Hence, the smallest inclusion set for y^* computable with the aid of our algorithm M must include *all of $R_M(f)$*; see figure 1.

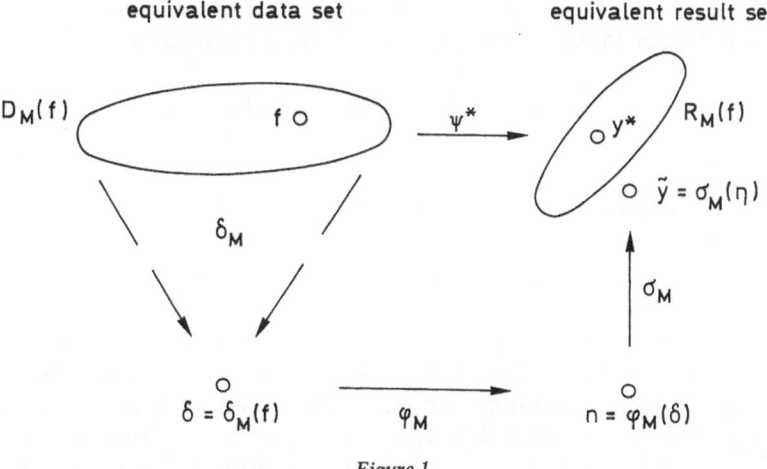

Figure 1

It may seem at first that so-called *adaptive methods* evade this situation by adapting their data value retrieval to the particular data function f. But they can only adapt to the values of f which they have so far retrieved: After the algorithm has finished, there exists a record of retrieval arguments and any function $g \in F$ which coincides with f at these arguments will be an *equivalent data function* and belong to $D_M(f)$. Thus adaptivity will only complicate the structure of $D_M(f)$ but not change the situation sketched in figure 1. Also if values of functionals other than function evaluation are retrieved by δ_M, the situation remains unchanged.

The extent of the dilemma is portrayed by the following

Proposition 1: Assume that we have a linear data space F and that

(i) all retrieved functionals d_v in δ_M are *linear*,
(ii) the set of conditions $d_v(g) = 0$, $v = 1(1)N$, has a *nontrivial solution* $g_0 \in F$, then $D_M(f)$ is unbounded.

Proof: $\delta_M(f + \gamma \cdot g_0) = \delta_M(f)$ for all $\gamma \in \mathbb{R}$. $\square$

In many cases — certainly if the mathematical problem is linear —, the unboundedness of $D_M(f)$ will imply the unboundedness of $R_M(f)$. But even if only a bounded part of D_M is in the domain of the solution mapping ψ^* (for some nonlinear problem), the resulting set $R_M(f)$ of equivalent results will tend to be large.

This means that, generally, *no guaranteed information* (errorbound, inclusion) can be *computed* if the hypotheses of Proposition 1 are satisfied.

Example (from numerical folklore):

$$\int_a^b f(t)\,dt \approx Q(f) = \sum_{v=1}^N \omega_v f(t_v).$$

In an adaptive quadrature algorithm, the t_v and N (and the ω_v) depend on the integrand f. Yet, if we run the algorithm for $f \equiv 0$ and record the t_v, we may have the algorithm produce $Q(g) = 0$ for any data function

$$g(t) = c \cdot \prod_{v=1}^N (t - t_v)^2, \qquad c \in \mathbb{R}.$$

while $\int_a^b g(t)\,dt$ may be given arbitrarily large values. $\square$

Within the framework of the algorithmic theory of Traub and Wozniakowsky, there arise a number of other situations where linear information turns out to be insufficient for more general computational procedures, not all as simple as the above one. E.g., Wasilowski [9] has shown that — in a strictly defined sense — no stationary iteration for the computation of a zero of a function f can be globally convergent if it uses only linear information about f. Note that Proposition 1 implies that — when an algorithm for the computation of an approximation for a zero has stopped after a finite number of evaluations of linear functionals of f — it is not possible to compute a bound for the error with the available information (or further linear information).

But linear information is all that one may expect to obtain from a standard, black-box function subroutine. Hence we have a first insight:

Input from (standard) function subroutines cannot suffice for automatic result verification.

This suggests two search directions for a sound basis for inclusion algorithms in Numerical Analysis:

(i) the use of *nonlinear* information,
(ii) a restriction of the space of *admissible data functions*.

3. Compact Equivalent Data Sets

Assume that — with a suitable topology in the data space F and in the result space R — the exact result mapping ψ^* for our problem is continuous. Then the *compactness* of the equivalent data set $D_M(f)$ is sufficient for the boundedness of the set $R_M(f)$ of equivalent results, which is in turn a prerequisite for the *existence* of inclusion algorithms.

The hypotheses used in conventional theorems about numerical methods indicate that functionals of the type

$$d(f) = \max_{t \in I} |f^{(p)}(t)| \tag{3.1}$$

serve our purpose of defining bounds for $R_M(t)$. There is, however, an essential difference between assuming such information in a theorem and using it in an algorithm:

It appears unrealistic to request that values of the type (3.1) should be directly retrievable from the subroutine for a data function. On the other hand, to request the specification of such a value by the user of the algorithm is as unrealistic, since few users will have any idea of the size of say the 5th derivative of a data function which they specify.

Furthermore, mere estimates of quantities like (3.1) will not suffice: If their values are specified too large the set of equivalent results and the related inclusion sets to be used for result verification will remain unduly large; if their values are too small, they may be inconsistent with the other data of f.

Example: One of the few algorithms which make direct use of compactifying data of the type (3.1) is Gaffney's inclusion algorithm for the range of a function [5]: Given

$$d_v(f) = f(t_v), \quad v = 1(1)N, \quad t_v \in [a, b], \tag{3.2}$$

$$\text{and} \quad d(f) = \max_{t \in [a, b]} |f^{(p)}(t)|, \quad \text{for some } p \le N, \tag{3.3}$$

the algorithm explicitly constructs the set $D(f)$ of equivalent functions, i.e. the set of all functions which will generate the same values for the d_v and d, by specifying the upper and lower boundary functions of that set, which turn out to be piecewise p-th degree polynomial functions.

In the specific case of figure 2 (from [5]) the specified bound for $|f^{(4)}|$ is obviously too generous for a reasonable inclusion of some quantity to be computed from f; on the other hand it excludes such candidates for $g \equiv f$ as the unique fifth degree interpolation polynomial whose 4-th derivative takes values in $[-191, 184]$.

Information on f:

$$f(t_v) \quad \text{at} \quad t_v = 1(1)6, \qquad \max_{t \in [1,6]} |f^{(4)}(t)| = 100.$$

t	1.0	2.0	3.0	4.0	5.0	6.0
$f(t)$	−1.0	1.0	6.0	0.0	3.0	−6.0

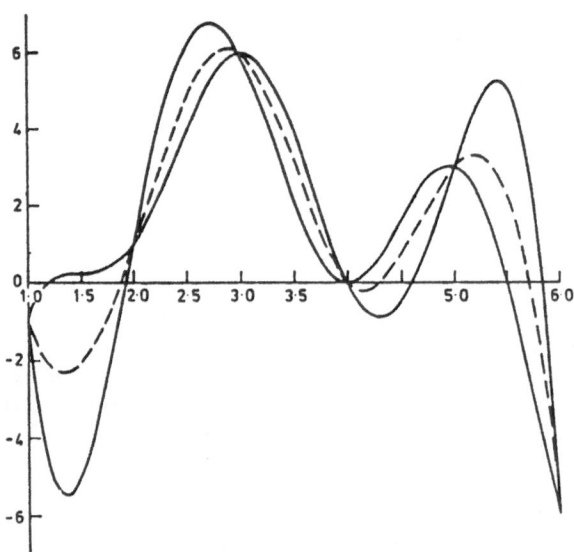

Figure 2

This example shows that reliance upon compactifying nonlinear information on f will generally not be a satisfactory basis for the design of inclusion algorithms.

4. Unique Specified Problem

In the framework of Section 2 (cf.e.g. figure 1), the characteristic distinction of *Numerical Algebra* is the biunique relation between the data of the mathematical problem and the data used by the computational algorithm. Thus the computation "knows all" about the specified mathematical problem, and the computation of arbitrarily good inclusions is a matter of algorithmic skill and of a suitable arithmetic.

This biunique relation makes it possible to define rigorously the mathematical problem — and hence its result — which is specified by a given set of data even if we are restricted to data in a system M of floating point numbers: For a given run of an implementation of an inclusion algorithm in a given computer arithmetic, the results to be enclosed are those of the respective mathematical problem with the data in $M \subset \mathbb{R}$ *as they are available to the object code.*

In the light of Sections 2 and 3, it appears a necessity to transfer this biunique relation between the data available to the object code of an inclusion algorithm and a *specified mathematical problem* also to *Numerical Analysis*. This requires that we consider only problems with data functions from a suitable subset D of F.

The choice of this data function sets D and the *computational representation of its elements* thus become the crucial preliminaries for the design of inclusion algorithms in Numerical Analysis. Here, flexibility and practicality play the primary role.

Example: The restriction of D to (piecewise) polynomials of a specified maximal degree may be desirable for the designer of inclusion algorithms; but it will not be sufficiently adaptable to frequently occurring needs: A function like $\exp(-t)$ cannot be well approximated by polynomials for $t > 0$.   □

A more flexible set has been proposed by N.J. Lehmann for the purposes under discussion (see e.g. [6]):

Example: Let D be the linear space of all solutions of all linear homogeneous o.d.e. with constant coefficients of a specified maximal order l. It is obvious that such a set D comprises exponential and trigonometric functions besides polynomials.

A particular $f \in D$ may be specified by $2l$ real numbers: l coefficients of the differential equations and l initial or boundary values; but the same f may also be specified by its l coefficients in a linear combination of suitable basis functions (which form an l-dimensional linear space). The availability of two specification modes makes the approach attractive; but the computational difficulties in non-trivial algorithms are considerable.   □

Instead of considering further natural choices of function sets D, let us list some *desirable properties* which the functions in D should have for the design of inclusion algorithms:

Their algebraic handling should be easy,
their accurate evaluation should be feasible (even with an "arbitrary" accuracy),
the formation of derivatives of such functions should be easy,
the tight bounding of their (derivative) values over domains should be feasible,
various analytic operations with them should be feasible,
their specification to an object code should be easy.

While it may seem at first that this list leads us back to sets of polynomials or closely related functions there exists a much more natural and powerful solution to our needs.

5. Arithmetic Expressions as Function Data

The following observation is crucial for an adequate specification of function data: Function data need not necessarily be specified by the usual function procedures (subroutines).

In a suitable software environment, a string like

$$'F:\ T \to 5.5 * \mathrm{EXP}(-T) + 2.4 * \mathrm{SIN}(Pi * T)' \tag{5.1}$$

should be a fully satisfactory specification of a function for a computational algorithm of Numerical Analysis.

First of all, the information consists of finitely many characters, which is a prerequisite from the computer science point of view. And quite naturally, the information permits the computation of approximate function values $f(t)$ for "all" t. It would even be possible to specify the requested accuracy of this approximate evaluation independently of the specification of f, e.g. in a dynamic accuracy environment. Also, this is precisely the class of functions whose evaluation results may be automatically verified.

Moreover, such a specification permits analytic and other functional operations as well:

The transition from the string (5.1) to the string

$$'FP1:\ T \to -5.5 * \mathrm{EXP}(-T) + 2.4 * Pi * \mathrm{COS}(Pi * T)'$$

may be achieved by a simple string manipulation procedure, which makes the *exact* derivative of f available as a new data function. Integration is also feasible in the case of (5.1), and — more important in the light of section 3 — the approximate evaluation as well as the a guaranteed enclosure of nonlinear functionals like (3.1) are possible.

Most important from our general point of view is the fact that there is a biunique relationship between the information in the string (5.1) and the mathematical object which is the represented function. Thus, function data presented in this form to the object code of some algorithm in Numerical Analysis *specify a unique mathematical problem*. It is now a matter of the skill of the algorithm designer to generate an arbitrarily tight approximation or inclusion of the result of this mathematical problem since all information about the specified problem is available to the code.

From this point of view, it appears natural to consider the following function set F_0 as the reservoir of *admissible data functions* from $\mathbb{R}$ to $\mathbb{R}$ in Numerical Analysis:

F_0 *contains all functions consisting of a finite[2] number of "pieces" each of which may be specified by an arithmetic expression[2] using the rational operations and a given set of standard functions and bracketing.*

Mathematically, this is a very rich set of functions if the standard functions include the usual set of exponential, logarithmic, and trigonometric functions and general powers. Although the finiteness[2] of the expressions will prevent that every continuous function on a finite interval may be approximated arbitrarily well by functions from F_0, the quality of the approximation will suffice for all practical purposes. In particular, it will be much better than what could be achieved with a fixed finite dimensional linear space of functions, e.g. polynomials of a fixed maximal degree.

At the same time, the use of such functions is very natural for the human user. Nearly all functional dependencies in the natural sciences and in engineering, but also in the economical and social sciences may be expressed with the aid of functions in F_0.

Also it is clear how analogous sets of functions in several real variables (or in complex variables) may be specified. In the case of several variables, the definition of the domains for piecewise functions will have to be suitably restricted; but otherwise no intrinsically new situations arise.

When we consider the few existing inclusion algorithms of Numerical Analysis (in our sense of the word), we find that they use arithmetic expressions in string format for the specification of function data throughout.

This is true for the only ACRITH routine which may be considered to belong to Numerical Analysis, viz. DNLSS which encloses the zeros of nonlinear systems consisting of arithmetic expressions, see [1]. It is also true of Lohner's sophisticated algorithm and code which encloses the solutions of ordinary initial value problems, in $\mathbb{R}^n$, cf. [7]. Similarly, the enclosing quadrature algorithm by G. Corliss [4] assumes the integrand specified by an arithmetic string. The Numerical Analysis papers in this volume also require that data functions are specified *analytically*, i.e. in a form like (5.1).

---

[2] Naturally, there must be specified, implementation dependent bounds for the number of pieces and the length and depth of the arithmetic expressions.

This evidence also strongly supports our claim that F_0 — with a suitable set of admissible standard functions in the arithmetic expressions — is the proper set for the specification of function data in algorithms of Numerical Analysis with automatic result verification.

6. Consequences

When we regard the handling of *arithmetic expression strings* in the inclusion codes mentioned above we find that the handling and processing procedures for the strings representing the function data are part of the code. It is clear, on the other hand, that this is an unacceptable state of the art:

If character strings representing arithmetic expressions are essential data in algorithms of Numerical Analysis, this data type must be a fundamental object in languages for scientific computation. Furthermore, the basic handling and processing of this data type must also be expressible in such programming languages through generic statements which tell the compiler to substitute the respective procedures into the object code.

So far language constructs for the processing of arithmetic expression strings have been available only in Computer Algebra Systems, like MACSYMA, Maple, etc. (for an overview of such systems, see, e.g., [3]).

Nearly all of these programming systems, which have been mainly designed for interactive use, contain syntactic means for the introduction of function data from F_0 in the form of character strings, for the execution of various analytic operations with such objects which result in new arithmetic expression strings, and also for the execution of some numerical operations on such objects like evaluation at a specified argument. On the other hand, these programming system generally lack algorithmic potential for the formulation of nontrivial Numerical Analysis algorithms, at least in a convenient form. It is true that REDUCE, e.g., permits the transformation of an arithmetic expression string into a Fortran subroutine; but this is only a partial solution to our general problem.

It is a pity that — even in numerically powerful languages like Fortran-SC (see [2]) — the specification and handling of arithmetic expression strings as principal objects of algorithms in Numerical Analysis has not been provided for. To focus the attention of Numerical Analysts and Computer Scientists onto this huge gap has been one of the main objectives for the preparation of this paper.

It should be a particular challenge for Computer Scientists to design efficient and practical language and compiler extensions for the use of arithmetic expression strings as part of a more powerful problem solving environment than it is provided at this time by either a language like Fortran-SC or an interactive system like Maple.

While the fundamental procedures for the processing and I/O of function data in the form of arithmetic expression strings would have to be part of the language, the language would also have to provide the necessary tools for the formulation of algorithms of a more sophisticated form. This would particularly concern the formulation of inclusion algorithms in Numerical Analysis. The specification of the most suitable set of basic procedures necessary for that purpose (as part of the language) is a research object for Numerical Analysts. One procedure whose presence is definitely required is the optimal (or very near optimal) inclusion of the range of a function from F_0 over a given interval.

Another class of procedures which should be part of such a scientific computation environment, consists of algorithmic procedures from numerical data (in the form of reals) to function data (in the form of arithmetic expression strings):

Example: A subroutine $LSEXPS(N, T, Y, M, F, R)$ could accept data points (t_i, y_i), $i = 1(1)N$, in N-vectors T and Y and compute their least squares approximation by an M-term exponential sum $\sum_{m=1}^{M} a_m \exp(b_m t)$, leaving a least squares residual R. The value assigned to the output parameter F by this routine would be a character string representing the computed exponential sum as an arithmetic expression. Thus the variable at this position could immediately by used as input variable in an *analytic* integration or differentiation routine.

7. Conclusions

At this time, Scientific Computation is in a state of rapid growth. This is due to the decreasing cost and increasing sophistication of the available hardware on the one hand and due to the exploding needs of all areas of the quantitative sciences and of engineering where computer simulation has largely replaced experimentation and has become one of the main research and design tools. While large parts of Scientific Computation may well live on the software tools presently available, at least as far as numerical procedures are concerned, there is a growing area where more refined numerical software could open up qualitatively new approaches. One such numerical tool is the automatic result verification for specified mathematical problems.

We have attempted to point out the — possibly only — approach to the specification of data functions of problems in Numerical Analysis which is both theoretically sound and practically feasible for this purpose: The restriction of data functions to the set F_0 of arithmetic expressions as defined in Section 5, and the syntatic specification of such data by appropriate character strings. It is clear that further analysis and practical experience will be needed to obtain a clear picture of the most effective implementation of this approach.

It is hoped that the necessary steps towards such an implementation will be taken in the very near future. This will lead to a boost in the design of inclusion codes in Numerical Analysis, an event which will be welcomed by many colleagues in the Applied Sciences.

References

[1] ACRITH (IBM High-Accuracy Arithmetic Subroutine Library), Program Description and User's Guide, SC 33-6164-02, 3rd Edition, 1986.

[2] Bleher, J. J., et al.: FORTRAN-SC, a study of a FORTRAN extension for engineering scientific computation with access to ACRITH, Computing 39 (1987), 93−110. Also this Volume.

[3] Buchberger, B., Collins, G. E., Loos, R., (Eds.): Computer Algebra − Symbolic and Algebraic Computation, Springer, Wien, New York (1983).

[4] Corliss, G. F., Rall, L. B.: Adaptive, selfvalidating quadrature, SIAM JSSC 8 (1987), 831−847.

[5] Gaffney, P. W.: To compute the optimal interpolation formula, Math. of Comp. 32 (1978), 763−777.

[6] Lehmann, N. J.: Die Analytische Maschine − Grundlagen einer Computer-Analytik, Sber. Sächs. Akad. Wiss. Leipzig, Math. Nat. Kl., 118 (1985), Heft 4.

[7] Lohner, R. J.: Enclosing the solution of ordinary initial and boundary value problems, in: Computer Arithmetic: Scientific Computation and Programming Languages, Eds.: Kaucher, E. W., Kulisch, U. W., Ullrich, Ch., Wiley-Teubner, Stuttgart (1987), 255−286.

[8] Traub, J. F., Wozniakowski, H.: A General Theory of Optimal Algorithms, Academic Press, New York (1980).

[9] Wasilowski, G. W.: Can any stationary iteration using linear information be globally convergent?, J. ACM 27 (1980), 263−269.

Prof. Dr. H. J. Stetter
Institut für Angewandte
und Numerische Mathematik
Technische Universität Wien
Wiedner Hauptstrasse 8–10
A-1040 Wien
Austria

Appendix

Computing, Suppl. 6, 227–244 (1988)

Computing
© by Springer-Verlag 1988

FORTRAN-SC

A Study of a FORTRAN Extension for Engineering/Scientific Computation with Access to ACRITH*

J. H. Bleher and S. M. Rump, Böblingen
U. Kulisch, M. Metzger, Ch. Ullrich and W. Walter, Karlsruhe

Received June 16, 1987; revised August 7, 1987

Abstract — Zusammenfassung

FORTRAN-SC. A Study of a FORTRAN Extension for Engineering/Scientific Computation with Access to ACRITH. A new programming language called FORTRAN-SC is presented which is closely related to FORTRAN 8x. FORTRAN-SC is a FORTRAN extension with emphasis on engineering and scientific computation. It is particularly suitable for the development of numerical algorithms which deliver highly accurate and automatically verified results. The language allows the declaration of functions with arbitrary result type, operator overloading and definition, as well as dynamic arrays. It provides a large number of predefined numerical data types and operators. Programming experiences with the existing compiler have been very encouraging. FORTRAN-SC greatly facilitates programming and in particular the use of the ACRITH subroutine library [14], [15].

Key words: Programming languages, FORTRAN, compiler, computer arithmetic, numerical computation, verified numerics.

FORTRAN-SC. Eine FORTRAN-Erweiterung für naturwissenschaftlich-technisches Rechnen mit Zugriff auf ACRITH. FORTRAN-SC ist eine neue Programmiersprache, welche mit FORTRAN 8x eng verwandt ist. Es handelt sich um eine FORTRAN-Erweiterung für Anwendungen im naturwissenschaftlich-technischen Bereich. Insbesondere eignet sich FORTRAN-SC für die Entwicklung von numerischen Algorithmen, welche hochgenaue und automatisch verifizierte Ergebnisse liefern. Die Sprache ermöglicht die Vereinbarung von Funktionen mit allgemeinem Ergebnistyp, das Überladen und Definieren von Operatoren, sowie dynamische Felder. Außerdem wird eine große Zahl vordefinierter numerischer Datentypen und Operatoren zur Verfügung gestellt. Die bisherigen Programmiererfahrungen mit dem existierenden Compiler sind sehr vielversprechend. FORTRAN-SC vereinfacht das Programmieren und insbesondere die Benutzung der ACRITH-Unterprogrammbibliothek wesentlich [14], [15].

1. Introduction

In electronic computers the elementary arithmetic operations are these days generally approximated by floating-point operations of highest accuracy. This is one of the essential intentions of the ANSI/IEEE Floating-Point Arithmetic Standard 754 [5]. This arithmetic standard also requires the four basic arithmetic operations

---

* This paper was already published in Computing 39, 93–110 (1987).

$+, -, *, /$ with directed roundings. A large number of processors is already on the market which provide these operations. So far, however, no common programming language allows access to these operations.

On the other hand, there is a noticeable shift from general purpose machines to vector processors and parallel computers in scientific computation. These so-called super-computers provide additional arithmetic operations such as "multiply and add", "accumulate" or "multiply and accumulate" [12]. All of these hardware operations should always deliver a result of highest accuracy for all possible combinations of data. As far as we know, no processor which fulfills this requirement is as yet available. From the point of view of programming, there exists no standard language which permits direct access to these operations.

There is a continuous effort to enhance the power of programming languages. New powerful languages like ADA have been designed, and the development of existing languages like FORTRAN is constantly in progress. However, since these languages still lack a precise definition of the arithmetic, the same program may produce different results on different processors.

During the development of FORTRAN 8x, proposals were made as to how real and complex operations with directed roundings, interval and complex interval arithmetic, and vector/matrix operations for all these types could be incorporated into that language [8], [9]. Many useful concepts which are necessary for their realization have been adopted in recent drafts of the proposed FORTRAN 8x standard. Such concepts are: derived data types, dynamic arrays, functions with arbitrary result type, operator overloading and definition, and modules.

In this paper, we refer to the programming language described by the latest draft of the proposed FORTRAN 8x standard simply as FORTRAN 8x [4].

In 1983, a group of scientists[1] worked out the first draft of a programming language which is closely related to FORTRAN 8x. In particular, it is a superset of FORTRAN 77 [3]. The new language was given the name FORTRAN-SC. In this paper, we will give an informal description of this language. We will also make some remarks on its current implementation. FORTRAN-SC pursues the same goals as PASCAL-SC [10], [17], another programming language for scientific computation which was designed and implemented at the Institute for Applied Mathematics at the University of Karlsruhe.

With respect to scientific computation the newly designed language surpasses FORTRAN 8x. For example, FORTRAN-SC provides more than 1000 predefined arithmetic operators for all kinds of numerical data types. All of these operators are required to deliver a result of at least 1 ulp accuracy. This means that the error is less than 1 *unit* in the *last place* (1 ulp). In other words, there is no floating-point number between the computed and the exact result. Since our interest is mainly directed towards scientific computation, several features of FORTRAN 8x which are of minor importance for that purpose were not incorporated into FORTRAN-SC.

---

[1] J. H. Bleher, H. Böhm, G. Bohlender, A. T. Gerlicher, K. Grüner, E. Kaucher, R. Klatte, U. W. Kulisch, W. L. Miranker, M. Neaga, S. M. Rump, Ch. Ullrich, J. Wolff von Gudenberg.

In FORTRAN-SC, the mathematical definition of the arithmetic is part of the language. This definition includes the vector/matrix operations. The elementary arithmetic operations $+, -, *, /$ with directed roundings are axiomatically defined and directly accessible in the language. All other arithmetic operations, in particular for intervals, are defined according to the rules of semimorphism [18].

During the process of implementation, the new language was further developed and completed. The compiler consists of a complete front end performing full analysis of the source language and a code generator. To achieve high portability, the code generator produces FORTRAN 77 code. The extensive runtime library contains the predefined operators and intrinsic functions of FORTRAN-SC. All implemented arithmetic operators are accurate to at least 1 ulp.

Programming experiences in FORTRAN-SC have been very encouraging. As a result of the operator notation in expressions for all arithmetic data types, programs become clearer and more concise. FORTRAN-SC programs are easy to read, write and understand.

One of the main goals of the development of FORTRAN-SC was to facilitate the use of the ACRITH subroutine library. FORTRAN-SC makes all ACRITH subroutines which provide arithmetic operations available as predefined operators. Furthermore, the large number of ACRITH mathematical standard functions for real, complex, interval and complex interval data may be referenced by their specific or generic names in FORTRAN-SC.

Through the availability of the interval, vector and matrix data types, the use of the ACRITH problem solving routines is greatly simplified. It is a common characteristic of all ACRITH routines that whenever they deliver a result, it is verified to be correct by the computer.

The language FORTRAN-SC was developed and implemented in a collaboration of the IBM Development Laboratory in Böblingen, Federal Republic of Germany, with the Institute for Applied Mathematics at the University of Karlsruhe.

2. The Language FORTRAN-SC

In traditional programming languages like FORTRAN, ALGOL or PASCAL, each vector/matrix operation such as matrix multiplication or vector addition requires an explicit loop construct or a call to an appropriate procedure. The same is true for all operations with a non-scalar (or structured) result, e.g. for interval arithmetic. To avoid such difficulties, the language FORTRAN-SC provides all vector and matrix operations in the commonly used linear spaces (the real and complex numbers, real and complex vectors, real and complex matrices as well as all the corresponding interval spaces) as predefined operators. All of these operations are accessible through their usual operator symbol. Thus, expressions of these types can be written in the usual mathematical notation.

Additionally, a general operator concept is available in FORTRAN-SC. It enables the user to define his own operators for old and new data types. Other modern programming languages like ADA and FORTRAN 8x also provide an operator concept and the possibility to write non-scalar expressions. Thus, all of these languages provide the necessary tools for writing expressions in the common linear spaces in mathematical notation. The difference is, however, that FORTRAN-SC requires all predefined operators to deliver results of 1 ulp accuracy for all possible combinations of the data. The implemented runtime library satisfies this requirement.

Moreover, FORTRAN-SC provides means for computing certain classes of vector and matrix expressions with 1 ulp accuracy. In contrast, vector/matrix operations and expressions evaluated in the traditional manner will not, in general, deliver high accuracy.

It is an essential idea of FORTRAN-SC that the user need not perform an error analysis for any basic vector/matrix operation provided by the language. As a matter of fact, this should be a natural requirement for any modern programming language. Whenever a language provides a higher-dimensional operation by an operator symbol, the result should be of 1 ulp accuracy (as required in FORTRAN-SC). Otherwise, an error message should be given.

2.1. Standard Data Types, Predefined Operators and Functions

The following scalar data types are available in FORTRAN-SC:

INTEGER, REAL, DOUBLE REAL, COMPLEX, DOUBLE COMPLEX, INTERVAL, DOUBLE INTERVAL, COMPLEX INTERVAL, DOUBLE COMPLEX INTERVAL, LOGICAL, CHARACTER.

For these basic data types, arrays can be declared in the usual manner. For a large number of numerical array types, operators are predefined which always deliver 1 ulp accuracy. This means that the traditional arithmetic operators $+, -, *, /$ are to be implemented with the rounding to the nearest machine-representable element. The newly introduced arithmetic operators $+\langle, -\langle, *\langle, /\langle$ and $+\rangle, -\rangle, *\rangle, /\rangle$ are to be implemented with the monotone downwardly directed and upwardly directed rounding, respectively. For real and double real scalar data the latter operations are also provided by the ANSI/IEEE Arithmetic Standard 754 [5].

Tables 1 and 2 show all predefined arithmetical and relational operators. We use the following abbreviations for the basic numerical data types:

$$
\begin{aligned}
&\text{R} - \text{REAL} \\
&\text{DR} - \text{DOUBLE REAL} \\
&\text{C} - \text{COMPLEX} \\
&\text{DC} - \text{DOUBLE COMPLEX} \\
&\text{I} - \text{INTERVAL} \\
&\text{DI} - \text{DOUBLE INTERVAL} \\
&\text{CI} - \text{COMPLEX INTERVAL} \\
&\text{DCI} - \text{DOUBLE COMPLEX INTERVAL}
\end{aligned}
$$

Table 1. *Predefined arithmetic operators*

right operand / left operand	INTEGER	R DR C DC	I DI / CI DCI	RVEC DRVEC CVEC DCVEC	IVEC DIVEC CIVEC DCIVEC	RMAT DRMAT CMAT DCMAT	IMAT DIMAT CIMAT DCIMAT
[1]	{+, −}	{+, −}	{+, −}	{+, −}	{+, −}	{+, −}	{+, −}
INTEGER	λ	Λ	λ	Π	{*}	Π	{*}
R DR C DC	Λ	Λ		Π		Π	
I DI / CI DCI	λ		Γ / ζ		{*}		{*}
RVEC DRVEC CVEC DCVEC	Θ	Θ		Ω			
IVEC DIVEC CIVEC DCIVEC	{*, /}		{*, /}		Φ		
RMAT DRMAT CMAT DCMAT	Θ	Θ		Π		Ω	
IMAT DIMAT CIMAT DCIMAT	{*, /}		{*, /}		{*}		Φ

[1] The operators in this row are monadic (i.e. no left operand).

$$\lambda := \{+, -, *, /, **\}$$
$$\Lambda := \{+, +\langle, +\rangle, -, -\langle, -\rangle, *, *\langle, *\rangle, /, /\langle, /\rangle, **\}$$
$$\Pi := \{*, *\langle, *\rangle\}$$
$$\Theta := \{*, *\langle, *\rangle, /, /\langle, /\rangle\}$$
$$\Omega := \{+, +\langle, +\rangle, -, -\langle, -\rangle, *, *\langle, *\rangle\}$$
$$\Phi := \{+, -, *, .IS., .CH.\}$$
$$\zeta := \{+, -, *, /, .IS., .CH.\}$$
$$\Gamma := \{+, -, *, /, .IS., .CH., **\}$$

.IS.: Intersection of two intervals
.CH.: Convex hull of two intervals

Table 2. *Predefined relational operators*

right operand / left operand	INTEGER	R DR	C DC	I DI CI DCI	RVEC DRVEC CVEC DCVEC	IVEC DIVEC CIVEC DCIVEC	RMAT DRMAT CMAT DCMAT	IMAT DIMAT CIMAT DCIMAT
INTEGER	Ψ	Ψ	Σ	Ξ				
R DR	Ψ	Ψ	Σ	Ξ				
C DC	Σ	Σ	Σ	Ξ				
I DI CI DCI				Δ				
RVEC DRVEC CVEC DCVEC					Σ	Ξ		
IVEC DIVEC CIVEC DCIVEC						Δ		
RMAT DRMAT CMAT DCMAT							Σ	Ξ
IMAT DIMAT CIMAT DCIMAT								Δ

$\Xi := \{.IN.\}$
$\Sigma := \{.EQ., .NE.\}$
$\Delta := \{.EQ., .NE., .SB., .SP., .DJ.\}$
$\Psi := \{.EQ., .NE., .LT., .LE., .GT., .GE.\}$

.SB.: Subset for two intervals
.SP.: Superset for two intervals
.DJ.: Disjoint intervals
.IN.: Membership of a point in an interval

The suffixes VEC and MAT are abbreviations for one-dimensional and two-dimensional arrays, respectively.

Tables 1 and 2 are very compact. For instance, the symbol Ω may be substituted by any operator listed in the set Ω. Furthermore, each operator of the set Ω may be applied to all type combinations listed in the corresponding rows and columns. So each occurrence of Ω in Table 1 represents 144 ($=9*4*4$) predefined operators.

Compared to FORTRAN 77, FORTRAN-SC provides an extended set of mathematical standard functions (see Table 3). All these functions are available for the basic data types real, complex, interval and complex interval in single and double precision. They can be referenced by their specific or their generic name. FORTRAN-SC requires the mathematical standard functions with a point result to be accurate to within 1 ulp. The interval functions must be accurate to within 2 ulps. In the implemented runtime library, the actual error bounds are usually only half as large. Only in rare cases will the error be slightly greater — but always within the prescribed bounds.

Table 3. *Mathematical standard functions*

	Function	Generic Name
1	Natural Logarithm	LOG
2	Common Logarithm	LOG 10
3	Exponential	EXP
4	Sine	SIN
5	Cosine	COS
6	Tangent	TAN
7	Cotangent	COT, COTAN
8	Arcsine	ASIN
9	Arccosine	ACOS
10	Arctangent	ATAN
11	Arccotangent	ACOT
12	Arctangent (x 1/x 2)	ATAN2
13	Hyperbolic Sine	SINH
14	Hyperbolic Cosine	COSH
15	Hyperbolic Tangent	TANH
16	Hyperbolic Cotangent	COTH
17	Inverse Hyperbolic Sine	ARSINH
18	Inverse Hyperbolic Cosine	ARCOSH
19	Inverse Hyperbolic Tangent	ARTANH
20	Inverse Hyperbolic Cotangent	ARCOTH
21	Square Root	SQRT
22	Square	SQR
23	Absolute Value	ABS
24	Argument of a Complex Number	ARG

Besides the mathematical standard functions, FORTRAN-SC provides all the necessary type transfer functions for conversion between the numerical data types. They exist for scalar and array types.

2.2. Dynamic Arrays

As an extension to FORTRAN 77, the concept of dynamic arrays is introduced. This greatly extends the capabilities supplied by conventional FORTRAN arrays, called static arrays.

Dynamic arrays provide the user with the capability of allocating or freeing storage space for an array during execution of a program. Thus, the same program may be used for arrays of any size without recompilation. Furthermore, storage space can be employed economically since only the arrays currently needed have to be kept in storage and since they always use exactly the space required in the current problem. Also, type compatibility and full storage access security are offered for dynamic arrays. Note that the concepts of assumed size arrays and adjustable arrays become obsolete. Dynamic arrays offer the same functionality while being much more versatile.

In FORTRAN 77, arrays whose dimensions are unknown a priori are implemented via pseudo-dynamic mechanisms. This means that a sufficently large work area must be provided by the main program to handle the pseudo-dynamic objects, like vectors and matrices.

These and many other disadvantages are avoided through the use of dynamic arrays. We list a few advantages of dynamic arrays:

— storage space used only when needed,
— array size may change during execution,
— no recompilation for arrays of different sizes,
— complete type and index checking,
— no extra arguments for array dimensions,
— no user-defined array workspace,
— no module space for dynamic array storage.

The DYNAMIC statement is used to declare named array types and/or to declare dynamic arrays.

An array type is characterized by the (scalar) data type of the array elements and the number of dimensions of the array. We call this information (i.e. element type and number of dimensions) the *array form* or simply the *form* of a (dynamic or static) array. Note that the size of an array is not part of this information.

An array form can be given a name or several distinct names, each identifying a different named array type. The type of a dynamic array may simply be specified as an array form, or it may be specified by an array type name.

Example: Declaration of named array types and dynamic arrays:

DYNAMIC/REAL (:)/A, B
DYNAMIC/VECTOR = REAL (:)/V, W, /MATRIX = COMPLEX (:, :)/
DYNAMIC/MATRIX/M, /POLYNOMIAL = REAL (:)/P, Q

These statements declare A, B, V, W, P, Q as real one-dimensional dynamic arrays and M as a complex two-dimensional dynamic array. Note that VECTOR and POLYNOMIAL are two different named array types even though they are used for arrays of the same form. Thus, A, V, and P all have different data types.

In order to obtain storage space for a dynamic array, an ALLOCATE statement can be executed which specifies the index range for each dimension of the array. The storage space of a dynamic array is deallocated by a FREE statement.

Example: Allocation and deallocation of dynamic arrays:

DYNAMIC/DOUBLEMATRIX = DOUBLE REAL (:, :)/ A, B, C
READ(*, *) I
ALLOCATE A, B (I : 2 * I, 10)
...
C = A + B
FREE A
ALLOCATE A (20, 20)

An existing (allocated) dynamic array may be reallocated by an ALLOCATE statement without prior execution of a FREE statement. Thus, in the above example, the FREE statement is optional. In this manner the same array variable can be changed in size during execution. Note that its contents are lost when doing this. Deallocating a non-allocated array has no effect.

Furthermore, allocation of a dynamic array occurs automatically when assigning the value of an array expression to a non-allocated array (e.g. in the statement C = A + B in the example above).

The storage of a dynamic array which is local to a subprogram is automatically released before control returns to the calling program unit unless the array name occurs in a SAVE statement. Obviously, a static array may neither be allocated nor deallocated.

Array inquiry functions facilitate the use of static and dynamic arrays. In particular, the functions LB and UB provide access to the lower and upper index bounds of an array.

2.3. Array-Valued Functions and User-Defined Operators

In most programming languages the result of a function has to be a single scalar value. In addition, FORTRAN-SC allows functions which return a dynamic array as result. Thus, the user is no longer forced to write a subroutine instead of an array-valued function.

This concept allows functions with a result array whose size is unknown to the calling program even at the time it is calling the function. In general, only the function itself knows the size of its result. It is therefore always the function's responsibility to allocate the dynamic result array. Of course, allocation of the result may be taken care of by array assignment inside the function (as in the example below).

The type of an array function is defined by declaring the function name like a dynamic array.

Example:

```
C    This function multiplies the real R with the vector W and
C    substracts the resulting vector from the vector V.
     FUNCTION RVFUN (R, W, V)
     REAL R
     DYNAMIC /REAL(:)/ V, W, RVFUN
     RVFUN = V − R * W
     END
```

In the calling program unit, the function name RVFUN must be declared as a real one-dimensional dynamic array in a DYNAMIC statement. In addition, the function *must* be declared to be an external routine. Thus, in the calling unit, the function name must appear in an EXTERNAL statement or in an OPERATOR statement as the implementing function of a user-defined operator.

For some applications it may be useful and more convenient to introduce operators. In FORTRAN-SC, an operator is defined by an operator symbol or name, the number and type(s) of its operand(s) and the implementing function. The OPERATOR statement is used to declare such user-defined operators. In this way, an external function with one or two arguments can be called as a monadic or dyadic operator, respectively.

In an expression, an operator is uniquely determined by the operator symbol or name, by its appearance as a monadic or dyadic operator, and by the type(s) of its operand(s).

Example: Definition and usage of an operator for the dyadic product of two real vectors:

```
     PROGRAM MAIN
     INTEGER DIM
     DYNAMIC /REAL(:)/ A, B, /REAL(:, :)/ C
     OPERATOR .MUL. = DYPROD (REAL(:), REAL(:)) REAL(:, :)
     READ(*, *) DIM
     ALLOCATE A, B (1 : DIM)
     ...
     C = A .MUL. B
     ...
     END
```

```
     FUNCTION DYPROD (X, Y)
     DYNAMIC /REAL(:)/ X, Y, /REAL(:, :)/ DYPROD
     ALLOCATE DYPROD (LB(X) : UB(X), LB(Y) : UB(Y))
     DO 10 i = LB(X), UB(X)
        DO 10 j = LB(Y), UB(Y)
10          DYPROD (i, j) = X(i) * Y(j)
     END
```

All standard operator symbols and names may be overloaded and/or redefined in this way. In the example above, if the operator symbol * were to be used instead of the user-defined operator name .MUL., then the predefined multiplication operator for two real vectors (the inner product) would no longer be accessible within the program unit MAIN.

Table 4 summarizes the intrinsic operator symbols and names and displays the priorities of all FORTRAN-SC operators. Note that the user is free to invent his own operator names (enclosed in periods as in FORTRAN 8x [4]).

Table 4. *Precedence of intrinsic and user-defined operators*

Priority		Operators
high	12	user-defined monadic operators
	11	**
	10	* / *〈 /〈 *〉 /〉 .IS.
	9	monadic + monadic −
	8	+ − +〈 −〈 +〉 −〉 .CH.
	7	//
	6	.LT. .LE. .EQ. .GE. .GT. .NE. .SB. .SP. .DJ. .IN.
	5	.NOT.
	4	.AND.
	3	.OR.
	2	.EQV. .NEQV.
low	1	user-defined dyadic operators

Overloading or redefining intrinsic operator symbols and names does not change their priority. Note that the operator priorities in Table 4 are the same as in FORTRAN 8x [4].

The possibility to introduce different named array types for the same array form allows the definition of different operators with the same operator symbol (or name) for operands of the same form.

Example: Replacing the DYNAMIC and the OPERATOR statement in program MAIN in the preceding example by

```
     DYNAMIC /COLUMN = REAL(:)/ A
     DYNAMIC /ROW = REAL(:)/ B, /REAL(:, :)/ C
     OPERATOR * = DYPROD (COLUMN, ROW) REAL (:, :)
```

will have the effect of overloading the operator ∗ for a new type combination. The standard multiplication operator for two real one-dimensional arrays (the inner product) will then still be accessible.

2.4. Evaluation of Expressions with High Accuracy

FORTRAN-SC provides a large number of predefined numerical operators and intrinsic functions. Although all of these primitives are highly accurate, expressions composed of several such elements do not necessarily yield results of high accuracy. However, techniques have been developed to evaluate numerical expressions with high and guaranteed accuracy.

A simple class of such expressions are the so-called dot product expressions. We distinguish three kinds which differ in their result form: scalar, vector and matrix dot product expressions. Each such expression consists of a sum where the terms are single elements of this form or single products which deliver results of this form. Examples of such expressions are:

$$s1 + s2 ∗ s3 - v1 ∗ v2 \qquad \text{of scalar form}$$
$$v1 + m1 ∗ v2 - s1 ∗ v3 \qquad \text{of vector form}$$
$$m1 - m2 ∗ m3 + s1 ∗ m4 \qquad \text{of matrix form}$$

where $s1, s2, s3$ are scalars, $v1, v2, v3$ are vectors and $m1, m2, m3, m4$ are matrices with matching dimensions. The element types may be REAL, DOUBLE REAL, COMPLEX and DOUBLE COMPLEX.

The language FORTRAN-SC provides a special notation which indicates that a dot product expression is to be evaluated with 1 ulp accuracy. To obtain the unrounded or correctly rounded result of a dot product expression, the user has to parenthesize the expression and precede it by the symbol # which may optionally be followed by a symbol for the rounding mode.

The possible rounding modes for dot product expressions are:

Symbol	Expression form	Rounding mode
#∗	scalar, vector or matrix	nearest
#⟨	scalar, vector or matrix	downwards
#⟩	scalar, vector or matrix	upwards
# #	scalar, vector or matrix	smallest enclosing interval
#	scalar only	exact, no rounding

In order to be able to store the unrounded result of a dot product expression, FORTRAN-SC provides the new data types DOT PRECISION and DOT PRECISION COMPLEX. Such results are produced by a dot product expression where no rounding is specified (see last row in the table above). The DOT PRECISION types are scalar data types of restricted accessibility. Variables of these types can only be added, subtracted and compared. They may appear as summands within any scalar dot product expression. A dot precision variable may only be assigned a dot precision value of the same type.

Example:

```
      DOT PRECISION D
      DYNAMIC /REAL(:)/ X, Y, Z, /REAL(:,:)/ A, B
      REAL R
      INTERVAL V
      READ (*,*) n
      ALLOCATE A(n,n), B (=A), X(n), Y, Z(=X)
      ...
      X = #*(Y−A*X)
      V = ##(X*Y−Y*Z+R)
      A = #*(A*B−B*A)
      ...
      D = #(∅)
      DO 1∅ j=1, n
          D=D+ #(A(j,j)*B(j,j))
1∅    CONTINUE
      R = #*(D)
      V = ##(D)
```

In practice, dot product expressions may contain a large number of terms, making an explicit notation very cumbersome. In mathematics the symbol Σ is used for short. For instance, if A_i, B_i are scalars or vectors or matrices for each $i = 1, \dots, k$, then the sum

$$\sum_{i=1}^{k} A_i B_i$$

represents a dot product expression. FORTRAN-SC provides the equivalent shorthand notation SUM for this purpose. In the example above, the last six lines could be replaced by:

```
      D = #(SUM (A(j,j)*B(j,j),j=1,n))
      R = #*(D)
      V = ##(D)
```

This shows that a result involving n multiplications and n−1 additions can be produced with a single rounding operation. In the last statement the exact dot product is rounded to the smallest possible interval enclosing the exact value of the expression. Thus, the bounds of the interval V will either be the same or two adjacent floating-point numbers.

Dot product expressions play a key role in numerical analysis. Iterative refinement or defect correction methods for linear and nonlinear problems usually lead to dot product expressions. Exact evaluation of these expressions eliminates cancellation.

Information that has been lost by rounding effects during an initial computation can often be recovered by defect correction. Such corrections can deliver results of full floating-point accuracy. In principle, there is no limit to the accuracy that can be obtained by these methods.

3. The Implementation of FORTRAN-SC

Since 1984, a FORTRAN-SC compiler has been developed for the IBM/370 architecture. First programming experiences have demonstrated the usefulness and effectiveness of the language and the reliability of the implementation.

The FORTRAN-SC compiler is essentially a 2-pass compiler. Its front end performs complete lexical, syntactical and semantical analysis of the source program. In order to achieve high portability, the code generator produces FORTRAN 77 code. For easy debugging, the FORTRAN-SC source code can optionally be merged as comments into the generated FORTRAN 77 code. The extensive runtime library provides the predefined operators, the intrinsic functions and some auxiliary routines (e.g. for array management). Error handling is integrated into every routine.

The guiding principle of FORTRAN-SC is to achieve higher accuracy and more reliable results in scientific computation. These ideas had a profound influence on both the language and its implementation. Several new concepts (new data types, dynamic arrays and dot product expressions) required new compilation techniques.

As mentioned earlier, FORTRAN-SC is closely related to FORTRAN 8x. In particular, it is a superset of FORTRAN 77. In contrast to FORTRAN 77, however, the current implementation of FORTRAN-SC does not support statement functions and entry statements (use separate routines instead), assumed size arrays and adjustable arrays (use dynamic arrays instead).

On vector machines, many runtime routines could be vectorized. In particular, the speed of array operations which work elementwise could be greatly increased. However, special care must be taken because the language FORTRAN-SC requires that all predefined operators deliver results of 1 ulp accuracy.

4. FORTRAN-SC Sample Program

The following program assumes that a function (APPINV) for the computation of an approximate inverse of a square matrix exists. After preliminary inversion, the solution of the linear system is enclosed in an interval vector by successive interval iterations. For details about this method, see [21].

Note that in FORTRAN-SC lower case letters are interpreted as upper case and that identifiers and operator names may be up to 31 characters in length.

```
      PROGRAM LINSYS
C     Verified solution of the linear system of equations
C        A · x = b
      DYNAMIC /REAL(:, :)/ A, R, UNIT, IDENTITY
      DYNAMIC /INTERVAL(:, :)/ E
      DYNAMIC /REAL(:)/ B
      DYNAMIC /INTERVAL(:)/ X, Y, Z
      INTEGER dim, i, j, iter
C     UNIT is an EXTERNAL function which delivers the identity
C     matrix of the given dimension
      EXTERNAL UNIT

C     The following operator declaration overloads the intrinsic operator .IN.
C     for a new operand type combination (2 interval vectors).
      OPERATOR .IN. = INCL (INTERVAL(:), INTERVAL(:)) LOGICAL
      OPERATOR .EXPAND. = EXPAND (INTERVAL(:)) INTERVAL(:)
C     APPINV is an EXTERNAL function which delivers an
C     approximate inverse of a real matrix
      OPERATOR .APPROXIMATE INVERSE. =
     &              APPINV (REAL(:, :)) REAL(:, :)
      WRITE(*, *) 'Please enter the dimension of the linear system'
      READ(*, *) dim
      ALLOCATE A(dim, dim), B(dim)
      WRITE(*, *) 'Please enter the matrix A'
      READ(*, *) ((A(i,j), j = 1, dim), i = 1, dim)
      WRITE(*, *) 'Please enter the right-hand side B'
      READ(*, *) (B(i), i = 1, dim)

      R = .APPROXIMATE INVERSE. A
C     R · b is an approximate solution of the linear system.
C     Z is a maximally accurate inclusion of R · b. It does not
C     usually include the true solution.
      Z = # #(R * B)
      IDENTITY = UNIT(dim)
C     A maximally accurate inclusion of I − R · A is computed.
      E = # #(IDENTITY − R * A)
      X = Z

      DO 20 iter = 1, 10
C     To obtain a true inclusion, the
C     interval vector X is slightly inflated.
          Y = .EXPAND. X
C     The following expression contains interval vectors and an
C     interval matrix.
          X = Z + E * Y
          IF (X .IN. Y) GOTO 10
20    CONTINUE
```

```
      WRITE (*, *) 'No solution found !'
      STOP
10    WRITE (*, *) 'The given matrix is non-singular and the',
      &              'solution of the linear system is contained in:'
      WRITE (*, *) X
      END


      FUNCTION EXPAND (X)
      DYNAMIC /INTERVAL (:)/ X, EXPAND
      INTERVAL IEPS
      INTEGER i
      DATA IEPS /-1.0D-75, 1.0D-75/
      ALLOCATE EXPAND (=X)
C     EXPAND now has the same index bounds as X.
      DO 10 i=LB (X), UB (X)
         EXPAND (i) = X (i) + IEPS
10    CONTINUE
      RETURN
      END


      FUNCTION INCL (X, Y)
C     Is X a subset of the interior of Y?
      LOGICAL INCL
      DYNAMIC /INTERVAL (:)/ X, Y
      INTEGER i
      INCL = .TRUE.
      DO 10 i=LB (X), UB (X)
         IF (INF (Y (i)) .GE. INF (X (i)) .OR.
      &      SUP (Y (i)) .LE. SUP (X (i))) THEN
            INCL = .FALSE.
            RETURN
         END IF
10    CONTINUE
      RETURN
      END
```

5. Conclusion

Several modern programming languages provide a large number of basic arithmetic operations by their usual mathematical symbol. ADA, FORTRAN 8x and other languages appropriate for vector machines provide vector and matrix operations. It is certainly the most natural requirement that these basic operations be executed with highest accuracy for all possible combinations of data. If this is not possible, an error message should be given. In FORTRAN-SC, all vector and matrix operations deliver a result of at least 1 ulp accuracy.

Finally, FORTRAN-SC greatly simplifies the use of the ACRITH library. ACRITH provides routines for a large number of vector and matrix operations as well as for elementary functions. All of these can be accessed by their usual mathematical notation in FORTRAN-SC. The availability of additional higher data types as well as dynamic arrays and array-valued functions provide additional advantages. The problem-solving routines of ACRITH and other libraries may be called with a reduced list of parameters. All of these concepts improve the readability of programs and facilitate debugging considerably.

References

[1] Agarwal, R. C., Cooley, J. W., Gustavson, F. G., Shearer, J. B., Slishman, G., Tuckerman, B.: New Scalar and Vector Elementary Functions for the IBM System/370. IBM Journal of Research and Development *30/2*, 126–144 (1986).

[2] Alefeld, G., Herzberger, J.: Introduction to Interval Analysis. New York: Academic Press 1983.

[3] American National Standards Institute: American National Standard Programming Language FORTRAN. ANSI X3.9-1978.

[4] American National Standards Institute: American National Standard Programming Language FORTRAN. Draft S8, Version 104, ANSI X3.9-198x (1987).

[5] American National Standards Institute / Institute of Electrical & Electronic Engineers: A Standard for Binary Floating-Point Arithmetic. ANSI/IEEE Std. 754-1985, New York (Aug. 1985).

[6] Arithmos (BS 2000) Benutzerhandbuch. Siemens, U2900-J-Z87-1 (Sept. 1986).

[7] Bohlender, G., Kaucher, E., Klatte, R., Kulisch, U., Miranker, W. L., Ullrich, Ch., Wolff v. Gudenberg, J.: FORTRAN for Contemporary Numerical Computation. IBM Research Report RC 8348 (1980). Computing *26*, 277–314 (1981).

[8] Bohlender, G., Böhm, H., Grüner, K., Kaucher, E., Klatte, R., Krämer, W., Kulisch, U., Rump, S. M., Ullrich, Ch., Wolff v. Gudenberg, J., Miranker, W. L.: Proposal for Arithmetic Specification in FORTRAN 8x. Proceedings of the International Conference on: Tools, Methods and Languages for Scientific and Engineering Computation, Paris 1983, North Holland (1984).

[9] Bohlender, G., Böhm, H., Braune, K., Grüner, K., Kaucher, E., Kirchner, R., Klatte, R., Krämer, W., Kulisch, U., Miranker, W. L., Ullrich, Ch., Wolff v. Gudenberg, J.: Application Module: Scientific Computation for FORTRAN 8x. Modified Proposal for Arithmetic Specification According to Guidelines of the X3J3-Meetings in Tulsa and Chapel Hill. Report of the Institute for Applied Mathematics, University of Karlsruhe (March 1983).

[10] Bohlender, G., Rall, L. B., Ullrich, Ch., Wolff v. Gudenberg, J.: PASCAL-SC: Wirkungsvoll programmieren, kontrolliert rechnen. Mannheim-Wien-Zürich: Bibliographisches Institut – Wissenschaftsverlag 1986.

[11] Braune, K., Kraemer, W.: High-Accuracy Standard Functions for Intervals. Computer Systems: Performance and Simulation (Ruschitzka, M., ed.). North-Holland (1986).

[12] Buchholz, W.: The IBM System/370 Vector Architecture. IBM Systems Journal *25/1* (1986).

[13] Gal, S.: Computing Elementary Functions: A New Approach for Achieving High Accuracy and Good Performance. IBM Technical Report 88.153 (1985).

[14] IBM High-Accuracy Arithmetic Subroutine Library (ACRITH): General Information Manual, GC 33-6163-02, 3rd Edition (April 1986).

[15] IBM High-Accuracy Arithmetic Subroutine Library (ACRITH): Program Description and User's Guide, SC 33-6164-02, 3rd Edition (April 1986).

[16] IBM System/370 RPQ: High-Accuracy Arithmetic. SA 22-7093-0 (1984).

[17] Kulisch, U. (ed.): PASCAL-SC: A PASCAL Extension for Scientific Computation. Information Manual and Floppy Disks, Version IBM PC/AT, Operating System DOS, B. G. Teubner, Stuttgart. Chichester: John Wiley & Sons 1987.

[18] Kulisch, U., Miranker, W. L.: Computer Arithmetic in Theory and Practice. New York: Academic Press 1981.

[19] Kulisch, U., Miranker, W. L. (eds.): A New Approach to Scientific Computation. New York:
     Academic Press 1983.
[20] Moore, R. E.: Interval Analysis. Englewood Cliffs, N. J.: Prentice Hall 1966.
[21] Rump, S. M.: Solving Algebraic Problems with High Accuracy. In: [19], pp. 58 – 62.

J. H. Bleher and S. M. Rump   U. Kulisch, M. Metzger,
Entwicklung und Forschung   Ch. Ullrich and W. Walter
IBM Deutschland GmbH.   Institut für Angewandte Mathematik
Schönaicher Strasse 220   Universität Karlsruhe
D-7030 Böblingen   Postfach 6980
Federal Republic of Germany   D-7500 Karlsruhe
                                       Federal Republic of Germany

Defect Correction Methods

Theory and Applications

Edited by **K. Böhmer** and **H. J. Stetter**

1984. 32 figures. IX, 243 pages.
Soft cover DM 72,–, öS 504,–
Reduced price for subscribers to "Computing":
Soft cover DM 64,80, öS 453,60
ISBN 3-211-81832-4

(Computing, Supplementum 5)

Defect Correction Methods comprise an important class of constructive mathematical methods, many of which have been developed within the past 10 years. This volume contains a collection of papers from the major areas where defect correction methods have been devised and applied, and an introductory survey. It originated from an Oberwolfach meeting in July 1983; the articles were written by an international group of scientists who are active in this field. The volume contains the first comprehensive presentation of this important area of numerical and applied mathematics, an area whose results have so far only been published in journals and reports.
The reader will get acquainted with the ideas of defect correction through major theoretical results and through a variety of applications. The articles relate defect correction with discretization methods of many kinds (e. g. the novel multigrid technique), with algorithms for the computation of guaranteed high-accuracy results, and with design techniques in numerical software. The lists of references of the individual articles provide an easy access to the current literature on the subject.

Springer-Verlag Wien New York

Moelkerbastei 5, A-1010 Wien · Heidelberger Platz 3, D-1000 Berlin 33 ·
175 Fifth Avenue, New York, NY 10010, USA ·
37-3, Hongo 3-chome, Bunkyo-ku, Tokyo 113, Japan

Computer Algebra
Symbolic and Algebraic Computation

Edited by
B. Buchberger, G. E. Collins, and **R. Loos,**
in cooperation with **R. Albrecht**

Second Edition
1983. 5 figures. VII, 283 pages.
Soft cover DM 64,–, öS 448,–
ISBN 3-211-81776-X

Computer algebra is an alternative and complement to numerical mathematics. Its importance is steadily increasing. This volume is the first systematic and complete treatment of computer algebra. It presents the basic problems of computer algebra and the best algorithms now known for their solution with their mathematical foundations, and complete references to the original literature. The volume follows a top-down structure proceeding from very high-level problems which will be well-motivated for most readers to problems whose solution is needed for solving the problems at the higher level. The volume is written as a supplementary text for a traditional algebra course or for a general algorithms course. It also provides the basis for an independent computer algebra course.

Springer-Verlag Wien New York
Moelkerbastei 5, A-1010 Wien · Heidelberger Platz 3, D-1000 Berlin 33 ·
175 Fifth Avenue, New York, NY 10010, USA ·
37-3, Hongo 3-chome, Bunkyo-ku, Tokyo 113, Japan